INTEGRATED FLOOD AND DROUGHT MITIGATION MEASURES AND STRATEGIES
Case Study: The Mun River Basin, Thailand

Saowanit Prabnakorn

Thesis committee

Promotor
Prof. Dr C.M.S. de Fraiture
Professor of Hydraulic Engineering for Land and Water Development
IHE Delft Institute for Water Education & Wageningen University & Research

Co-promotor
Dr F.X. Suryadi
Senior Lecturer in Land and Water Development
IHE Delft Institute for Water Education

Other members

Prof.Dr S.E.A.T.M. van der Zee, Wageningen University & Research
Prof.Dr M. Kok, TUDelft
Prof.Dr M.E. McClain, IHE Delft Institute for Water Education & TUDelft
Dr S. Weesakul, Hydro-Informatics Institute, Ministry of Science and Technology, Thailand

This research was conducted under the auspices of the SENSE Research School for Socio-Economic and Natural Sciences of the Environment

INTEGRATED FLOOD AND DROUGHT MITIGATION MEASURES AND STRATEGIES

AND STRATEGIES

Case Study: The Mun River Basin, Thailand

Thesis
submitted in fulfilment of the requirements of
the Academic Board of Wageningen University and
the Academic Board of the IHE Delft Institute for Water Education
for the degree of doctor
to be defended in public
on Wednesday, 22 January 2020, at 3:30 p.m.
in Delft, the Netherlands

by

Saowanit Prabnakorn
Born in Bangkok, Thailand

Published by:
CRC Press/Balkema
Schipholweg 107C, 2316 XC, Leiden, the Netherlands
Pub.NL@taylorandfrancis.com
www.crcpress.com – www.taylorandfrancis.com

ISBN: 978-0-367-90378-7 (Taylor & Francis Group)
ISBN: 978-94-6395-136-4 (Wageningen University)
DOI: https://doi.org/10.18174/501426

To my beloved father and my dear family

Table of content

Thesis summary

Floods and droughts are the two major natural disasters that create significant impacts on people lives and their well-being. Floods occurred most frequently with the largest number of people affected. While droughts were less witnessed; however, they caused devastating impacts on human lives. Both catastrophes are expected to happen more frequently with a higher degree of severity due to the influence of climate change.

Both events can arise in the same river basin at different periods, as found in many river basins. For example, the Mun River Basin in Thailand, which is selected as the study area. The basin situated in the northeast, where the most extensive areas of rice cultivation located. Approximately 90% of rice cultivation here is rain-fed, and it is susceptible to flooding in the wet season and droughts in the dry season. As a result, rice yield here is the lowest in the country, and that makes many farmers persist in poverty.

This attracts attention from all stakeholders such as the government, authorities, farmers, etc., to tackle both hazards. However, due to differences in their natural characteristics and time of occurrences, assessments of floods and droughts as well as decision-making associated with mitigation measures, management plans and policies are usually performed and implemented separately. This may overlook opportunities to capture, store, and use the floodwater later when there is a water shortage. Moreover, this may cause overlap and inconsistency in operations and management, which costs more money.

Therefore, the study aims to assess floods and droughts and their impacts on agriculture at the basin scale and attempt to tackle them simultaneously by using integrated measures and strategies. The research is conducted at the Mun River Basin in Thailand, focusing on rice cultivation, which is an important contributor to the Thai economy.

The study begins with the investigation of climate variability, trends, and their impacts on rice production. Standardized Precipitation and Evapotranspiration Index at 1-month timescale (SPEI-1) is also included as one of the variables because it is based on both precipitation and temperature. The results reveal that the trends of minimum and maximum temperatures, precipitation, and SPEI-1 are predominantly increasing. The upward trends in minimum and maximum temperatures produce modest yield losses. While, the increasing trends in precipitation and SPEI-1 commonly benefit rice yields in all rice-growing months, except in the wettest month (mostly September). Overall, the total yield losses due to recent climate trends are rather low (less than 3% of the actual average yield). However, if the rising trends in minimum and maximum temperatures persist, there is a high possibility that the yield losses will become more severe in the future.

Next, the identification of flood and drought-prone areas is executed. The inundated areas at 10, 25, 50, and 100-year return period are obtained from an integrated hydrologic (SWAT) and hydraulic (HEC-RAS) model. The flooded areas are larger on the left bank than the right and that the flood depths vary from 0-4 m. About 60% of the floodplain inundation is less than 1 m and mostly found at the upstream and central areas of the river. The flat terrains make flooding duration last long, causing damage to crop growth and yields, particularly rice, which occupies a majority of the area and mostly locates on the river banks.

Drought-prone areas are detected by using the proposed comprehensive drought risk assessment scheme that incorporates hazard, exposure, and vulnerability. The hazard is estimated from water deficits calculated with respect to rice water requirement. Exposure is based on population and rice field characteristics, while other physical and socioeconomic factors and coping capacity are used in the estimation of vulnerability. The findings demonstrate that drought hazard is most critical in October, which can reduce rice yields significantly. Rice fields in the central part are more exposed to droughts than in other areas. Extensive land is under high and moderate vulnerability. The higher drought risks emerge in October and November and decrease from north to south.

After flood and drought problems are identified, a coping capacity of existing and on-going water resources development projects are evaluated. The total storage capacity of those projects is sufficient to cope with both hazards, but the performance is inefficient and ineffective. Thus, several pieces of advice are provided to improve the performance of in-situ measures. Moreover, other measures are proposed to complement the existing measures and make the storage system more flexibility. The suggested measures are practicable, applicable, economical, and less environmental impacts. They have the potential to sustainably solve both floods and droughts in order to enhance rice production in the basin.

This study provides all dimensions associated with floods and droughts, their impacts on rice yields, and mitigation measures to tackle both hazards simultaneously and sustainably. The achievement over the two climate extremes will ensure adequate food availability and alleviate poverty in the basin. Furthermore, the study shows that a holistic approach to simultaneously solving both problems is efficiency as all drops of water are utilized to benefit agriculture, the primary sector that feeds a growing population.

1

General introduction

1.1 Floods and droughts

Floods and droughts are the two main meteorological catastrophes. Flooding occurs the most (33%), whereas droughts are less frequent (5%- Table 1-1), but cause more fatal impacts in terms of the number of the death toll. Combined, the hazards account for 80% of the total number of people affected by natural disasters (Table 1-2). Furthermore, they have created enormous environmental, social, and economic impacts, such as degraded wetland areas, diminished natural biodiversity, destroyed agricultural land and industrial facilities, and high economic costs. It is expected that climate change effects (i.e., more intense rainfall, increasing seawater levels, etc.) will exacerbate flood occurrences and impacts in the future (Jones, 1999; quoted Lehner, Döll, Alcamo, Henrichs, & Kaspar, 2006). In addition to increased climate variability, population growth and societal interference are the other critical factors that exacerbate adverse consequences (Changnon, Pielke Jr, Changnon, Sylves, & Pulwarty, 2000; Douben, 2006; Kuntiyawichai, 2012; Pielke Jr & Downton, 2000).

Table 1-1   Number of natural disasters between 1900 and 2014.

Disaster Type	Continent					Worldwide	
	Africa	Americas	Asia	Europe	Oceania	Total	%
Epidemic	779	163	339	49	19	1,349	10%
Insect infestation	68	3	11	1	2	85	1%
Drought	293	137	153	42	22	647	5%
Extreme temperature	11	110	151	226	6	504	4%
Wildfire	28	136	82	99	36	381	3%
Earthquake	79	274	692	162	53	1,260	10%
Mass movement dry	5	18	21	9	2	55	0%
Mass movement wet	31	165	342	69	18	625	5%
Volcano	17	82	97	12	23	231	2%
Flood	868	1,000	1,775	533	133	4,309	33%
Storm	229	1,198	1,527	453	289	3,696	28%
Total	2,408	3,286	5,190	1,655	603	13,142	100%
Total (%)	18%	25%	39%	13%	5%	100%	

Source: EM-DAT: The OFDA/CRED International Disaster Database; www.em-dat.net - Université Catholique de Louvain - Brussels - Belgium, 2014.

Table 1-2   Number and percentage (in the parentheses) of victims and damage from natural disasters accounted for between 1900 and 2014.

Disaster Type	Number of deaths (10^6 people)	Number of injured (10^6 people)	Number of affected (10^6 people)	Number of homeless (10^6 people)	Total affected (10^6 people)	Total damage (10^9 USD)
Epidemic	10	0	45	0	45 (0.6)	0 (0)
Insect infestation	0	0	0.5	0	0.5 (0.01)	0.2 (0.01)

Disaster Type	Number of deaths (10⁶ people)	Number of injured (10⁶ people)	Number of affected (10⁶ people)	Number of homeless (10⁶ people)	Total affected (10⁶ people)	Total damage (10⁹ USD)
Drought	12	0	2,173	0	2,173 (31)	133 (5)
Extreme temperature	0	2	97	0	99 (1.4)	62 (2)
Wildfire	0	0	6	0	6 (0.1)	54 (2)
Earthquake	3	3	156	23	181 (3)	764 (29)
Mass movement dry	0	0	0	0	0 (0)	0.2 (0.01)
Mass movement wet	0	0	9	4	14 (0.2)	9 (0.3)
Volcano	0	0	5	0	6 (0.1)	3 (0.1)
Flood	7	1	3,469	89	3,559 (50)	639 (24)
Storm	1	1	920	53	974 (14)	986 (37)
Total	33	8	6,880	169	7,057 (100)	2,649 (100)

Source: EM-DAT: The OFDA/CRED International Disaster Database; www.em-dat.net - Université Catholique de Louvain - Brussels - Belgium, 2014.

1.1.1 Flooding

From 1900-2014, flooding has been a dominant disaster, accounting approximately for one-third of total natural disaster occurrences worldwide (Table 1-1). Table 1-3 presents that approximate 40% of all flood events took place in Asia, resulting in the largest number of deaths and people affected of any continent. Oceania, however, experienced very few flood events and had the lowest proportion of people affected. For damage losses, Asia had the highest damage, followed by Europe and America. The losses in Europe were higher than those in America, although about half as many floods happened. This is partly due to changes in land use. The severe 2011 flood of Thailand had an estimated USD 40 billion of damage, and approximately 9.5 million victims (Guha-Sapir, Hoyois, & Below, 2013). This flood emphasizes the significant role of land-use changes (specificially urbanization), population growth, and demographic shifts in increasing flood losses. Therefore, those factors, together with rising trends of flood events (particularly in Asia, Figure 1-1),will increase the likelihood and the degree of losses if no appropriate flood protection measure is applied.

Table 1-3 Number and percentage (in the parentheses) of victims and damage from flooding between 1900 and 2014.

Continent	Number of flood events	Number of deaths (10⁶ people)	Total affected (10⁶ people)	Total damage (10⁹ USD)
Africa	868 (20)	0.03 (0.4)	68 (1.9)	7.5 (1.2)
Americas	1,000 (23)	0.1 (1.5)	88 (2.5)	106 (17)
Asia	1,775 (41)	7 (98)	3,387 (95)	383 (60)
Europe	533 (12)	0.01 (0.1)	15 (0.4)	127 (20)
Oceania	133 (3)	0.001 (0.01)	1.2 (0.03)	14.5 (2)

Continent	Number of flood events	Number of deaths (10⁶ people)	Total affected (10⁶ people)	Total damage (10⁹ USD)
Total	4,309 (100)	7 (100)	3,559 (100)	639 (100)

Source: EM-DAT: The OFDA/CRED International Disaster Database; www.em-dat.net - Université
  Catholique de Louvain - Brussels - Belgium, 2014.

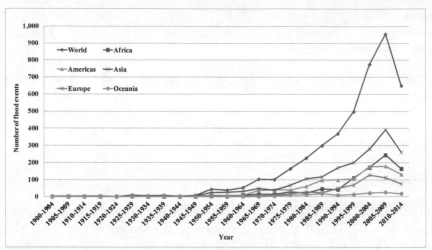

Source: EM-DAT: The OFDA/CRED International Disaster Database; www.em-dat.net -
  Université Catholique de Louvain - Brussels - Belgium, 2014.

Figure 1-1 Number of flood events by continent for 1980–2014.

1.1.2 Drought

Drought is the most damaging environmental phenomenon, and can cover large areas,
with more than one-half of all land susceptible to drought each year (Kogan, 1997). It
has occurred not only in arid and drought-prone areas, but also in regions with ample
precipitation (Pereira, Cordery, & Iacovides, 2002). Globally, the number of drought
events is minimal (Table 1-1). However, their adverse consequences are fatal as they
causes the highest number of deaths, and the seond most number of affected people
after flooding (Table 1-2). The vast majority of people affected were in Asia, followed
by Africa (where the most droughts occurred). The 1.2 million deaths in Europe
resulted from the 1921 drought event in the Soviet Union, and no mortality was
observed in other European countries in Europe since then. The greatest economic
damage was in America, followed by Asia and Europe, respectively (Table 1-4). In
addition, the occurrence of drought flucturates more than flooding, and they are
difficult to predict (Pereira et al., 2002). However, the drought trend is increasing
(Figure 1-2), showing the need for appropriate water management to secure water
availability.

Table 1-4 Number and percentage (in the parentheses) of victims and damage from droughts between 1900 and 2014.

Continent	Number of droughts	Number of deaths (10^6 people)	Total affected (10^6 people)	Total damage (10^9 USD)
Africa	293 (45)	0.85 (7)	367 (17)	3 (2)
Americas	137 (21)	0 (0)	70 (3)	59 (44)
Asia	153 (24)	10 (82.5)	1,713 (79)	34 (26)
Europe	42 (6)	1.2 (10)	15 (0.7)	25 (19)
Oceania	22 (3)	0.001 (0.01)	8 (0.4)	12 (9)
Total	647 (100)	12 (100)	2,173 (100)	133 (100)

Source: EM-DAT: The OFDA/CRED International Disaster Database; www.em-dat.net - Université Catholique de Louvain - Brussels - Belgium, 2014.

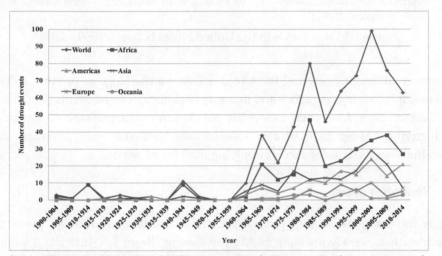

Source: EM-DAT: The OFDA/CRED International Disaster Database; www.em-dat.net - Université Catholique de Louvain - Brussels - Belgium, 2014.

Figure 1-2 Number of drought events by continent for 1980–2014.

1.2 Impacts on agriculture

Besides the detrimental effects of floods and droughts on human lives and properties, economic sectors such as agriculture can be badly affected by both calamities. Intense downpours and droughts create different impacts on the development of crops over short-term durations; however, crop growth will be eventually ceased if the situations last for a long time. As a result of floods and droughts, fewer yields are produced; for instance, corn production in the U.S. has been shown to reduce dramatically when floods and droughts occurred (Figure 1-3).

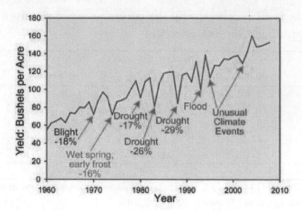

Source: Global Climate Change Impacts in the United States, U.S. Global Change Research
Program, 2009.

Figure 1-3 Corn yields in the United States between 1960 – 2008.

Therefore, all countries around the world will face a challenge to produce sufficient
food under increasing frequency and intensity of climate extreme events (Karl, Melillo,
& Peterson, 2009). Moreover, increasing population exerts pressure on agricultural
productivity. Owing to that, this research will primarily pay attention to the
agricultural sector, a major sector supplying food for all nations.

1.3 Research gaps

Floods and droughts can take place in the same river basin at different periods. For
example, they were witnessed in Murray-Darling Basin in Australia (Adamson,
Mallawaarachchi, & Quiggin, 2009), Thames Basin in the UK (WWF-UK, 2008), and
Chi and Mun river basins in Thailand (Chitradon, Boonya-aroonnet, &
Thanapakpawin, 2009). This can pose a significant challenge for water resources
management in these areas.

Long periods of excessive water on agricultural land caused by flooding can cause
damage to crop farms. Under such circumstances, floodwater is considered a threat to
plants, especially when it lasts over a long period. How to get rid of the excess water
as quickly as possible becomes the primary concern of stakeholders, such as the
government, civil servants, farmers, etc. They may overlook or do not take into account
opportunities to utilize the floodwater for other benefits. For example, it can be stored
and used later for agriculture in dry spells. It can also recharge groundwater by which
there will be more groundwater available for abstraction when it is needed (Pavelic et
al., 2012; UNEP-IETC, 1998).

Droughts also affect agricultural land due to insufficient water for plants to grow and mature. This also leads to conflict between farmers and other sectoral water users (Pavelic et al., 2012). Under such conditions in the dry season, those groups of authorities endeavor to seek out other sources of water in order to increase the water supply.

Furthermore, owing to differences in their inherent characteristics and time of occurrences, flood and drought analyses, mitigation measures assessments and management plans and strategies have been usually executed and implemented separately. This results in overlap and inconsistency in operations and management, which can be inefficient. In addition, the 21st century is predicted to be an era of water scarcity (Kundzewicz et al., 2002); thus all drops of water, even floodwater, should be used efficiently in spite of generally emptying into sea.

Therefore, a holistic approach which strives to assess and resolve floods and droughts simultaneously, by using integrated measures and strategies, is applied here. This may be a better alternative than the traditional approach that tackles them separately.

1.4 Selection of the study area and scope of the study

The study area is selected based on three criteria: 1) both floods and droughts occurred in the area, 2) agriculture is the most important sector contributing to the well-being of a majority of people and 3) no public study regarding floods and droughts has been publicly reported. Following those criteria, the Mun River Basin in northeast Thailand is chosen. The basin regularly experiences floods in the wet season and droughts in the dry season. For example, they both occurred (in different periods), in the years 2004, 2007-2016 (Department of Disaster Prevention and Mitigation, 2016; Department of Water Resources of Thailand, 2016; Hydro - Informatics Institute, 2019). Agriculture is the most affected sector because it occupies a considerable area of the basin. Rainfed-rice is a staple crop mostly found in lowland areas in the basin, so it is vulnerable to flooding and water scarcity. Thus, the scope of the study is specifically confined to floods and droughts, and their impacts on rice cultivation in the Mun River Basin in Thailand.

1.5 Research objectives

A large majority of rice fields in the basin are rainfed, making them susceptible to floods and droughts. When these hazards occur, they reduce yields or, in the worst case, cause crop failure. As a result, farmers may endure malnutrition, since rice cultivation here is primarily for self-consumption in families, and surpluses are sold when available (Haefele et al., 2006). Some may face poverty as farmers here have the lowest farm income in the country (Floch, Molle, & Loiskandl, 2007). Their income per capita is also the lowest compared to other regions in Thailand (Office of the National Economic and Social Development Board, 2014). Tackling floods and droughts in this basin will alleviate poverty and improve people's standard of living. Thus, based on this description, the main objective of this thesis is:

To assess flood and drought impacts on rice cultivation at the basin scale and attempt to tackle them simultaneously, by using integrated measures and strategies. Case study: the Mun River Basin in Thailand.

A primary research question and five sub-questions are formulated to facilitate the accomplishment of the research.

The primary research question is:

What are the extents of flood and drought impacts on rice cultivation in the basin and how to tackle them simultaneously?

Sub-questions are:

I.   How do the climatic variables change, and how does that impact rice growth and yields?

II.   Which rice fields are prone to flooding?

III. Which rice fields are susceptible to droughts?

IV.   What and where are water resources development projects currently implemented in the basin, and is the coping capacity sufficient to cope with both hazards?

V.   What are the integrated measures and strategies that could be potentially implemented to cope with floods and droughts in the basin?

1.6 Outline of the thesis

It is apparent that the study incorporates many aspects: climatic variation and trends, floods, droughts, rice cultivation, mitigation measures, and strategies. Thus, there is a large quantity of diverse data required in the analysis. The methodologies to deal with each part are also different and consist of many details. Moreover, the findings from each step demonstrate various aspects essential to fulfill all specific objectives, which eventually lead to achieving the main objective. All these elements are organized into seven chapters as follow:

- **Chapter 1** provides an introduction, background context and the rationale of the research topics

- **Chapter 2** describes the study area – the Mun River Basin, including physical characteristics, and social and economic status of people in the basin.

- **Chapter 3** reports the data and methodology used in the analysis of variations of climatic parameters, and their impacts on rice yields. The findings from this chapter highlight how climatic variables (such as minimum temperature, maximum temperature and precipition) impact rice yields at the basin scale over the past 30 years, and the extent to which they do this. Furthermore, the analysis includes Standardized Precipitation and Evapotranspiration Index (SPEI) as one of the variables because it is based on both precipitation and temperature.
  This chapter answers sub-question I.

- **Chapter 4** reports the data and methodology used in the identification of areas (rice fields in particular) inundated by flood hazard at different probabilities. The assessment involves the development of hydrologic and hydraulic models, including SWAT, SWAT-CUP, HEC-GeoRAS, HEC-RAS, and HEC-DSS. Also, rainfall frequency analysis is performed to derive rainfall depths at 10, 25, 50, and 100-year return periods.
  This chapter answers sub-question II.

- **Chapter 5** characterizes spatial variations in drought hazard, exposure, vulnerability, and risk for rice cultivation. A new scheme of drought risk assessment is proposed, which is based on the three key determinants of drought risk (hazard, exposure, and vulnerability). Following the scheme, physical, social, and economic data are needed. The analysis is carried out at monthly time steps from July to November (the rice-growing season) in order to understand drought impacts on rice growth at each stage.
  This chapter answers sub-question III.

- **Chapter 6** evaluates the coping capacities, performance, and factors influencing the achievement of the in-situ measures, and attempts to solve flood and drought problems simultaneously using the integrated mitigation measures and strategies.
  This chapter answers sub-question IV and V.

- **Chapter 7** provides a synthesis and recommends further research.

References

Adamson, D., Mallawaarachchi, T., & Quiggin, J. (2009). Declining inflows and more frequent droughts in the Murray–Darling Basin: climate change, impacts and adaptation*. *Australian Journal of Agricultural and Resource Economics, 53*(3), 345-366.

Changnon, S. A., Pielke Jr, R. A., Changnon, D., Sylves, R. T., & Pulwarty, R. (2000). Human Factors Explain the Increased Losses from Weather and Climate Extremes. *Bulletin of the American Meteorological Society, 81*(3), 437-442.

Chitradon, R., Boonya-aroonnet, S., & Thanapakpawin, P. (2009). Risk management of water resources in Thailand in the face of climate change. *Sasin Journal of Management*, 64-73.

Department of Disaster Prevention and Mitigation. (2016). *Disaster event report 2016 (in Thai)*. Retrieved from http://www.disaster.go.th/th/index.php

Department of Water Resources of Thailand. (2016). *Summary of the results of drought prevention and mitigation year 2015-2016. Final Report (in Thai)*. Bangkok, Thailand.

Douben, K.-J. (2006). Characteristics of river floods and flooding: a global overview, 1985–2003. *Irrigation and Drainage, 55*(S1), S9-S21. doi:10.1002/ird.239

Floch, P., Molle, F., & Loiskandl, W. (2007). Marshalling water resources: a chronology of irrigation development in the Chi-Mun River Basin, Northeast Thailand. *Colombo, Sri Lanka: CGIAR Challenge Program on Water and Food*.

Guha-Sapir, D., Hoyois, P., & Below, R. (2013). *Annual Disaster Statistical Review 2012: The numbers and trends*. In (pp. 50). Retrieved from http://www.emdat.be/

Haefele, S., Naklang, K., Harnpichitvitaya, D., Jearakongman, S., Skulkhu, E., Romyen, P., . . . Khunthasuvon, S. (2006). Factors affecting rice yield and fertilizer response in rainfed lowlands of northeast Thailand. *Field crops research, 98*(1), 39-51.

Hydro - Informatics Institute. (2019). Water events records (in Thai). Retrieved from http://www.thaiwater.net/v3/archive

Jones, J. (1999). Climate change and sustainable water resources: placing the threat of global warming in perspective. *Hydrological Sciences Journal, 44*(4), 541-557.

Karl, T. R., Melillo, J. M., & Peterson, T. C. (2009). *Global climate change impacts in the United States*: Cambridge University Press.

Kogan, F. N. (1997). Global drought watch from space. *Bulletin of the American Meteorological Society, 78*(4), 621-636.

Kundzewicz, Z. W., Budhakooncharoen, S., Bronstert, A., Hoff, H., Lettenmaier, D., Menzel, L., & Schulze, R. (2002). Coping with variability and change: Floods and droughts. *Natural Resources Forum, 26*(4), 263-274. doi:10.1111/1477-8947.00029

Kuntiyawichai, K. (2012). *Interactions between Land Use and Flood Management in the Chi River Basin.* (Ph.D. dissertation), UNESCO-IHE Institute of Water Education and Wageningen University, Delft, the Netherlands.

Lehner, B., Döll, P., Alcamo, J., Henrichs, T., & Kaspar, F. (2006). Estimating the impact of global change on flood and drought risks in Europe: a continental, integrated analysis. *Climatic Change, 75*(3), 273-299.

Office of the National Economic and Social Development Board. (2014). Per Capita Income of Population by Region and Province: 2005 - 2014. from Office of the National Economic and Social Development Board, Office of the Prime Minister

Pavelic, P., Srisuk, K., Saraphirom, P., Nadee, S., Pholkern, K., Chusanathas, S., . . . Smakhtin, V. (2012). Balancing-out floods and droughts: opportunities to utilize floodwater harvesting and groundwater storage for agricultural development in Thailand. *Journal of Hydrology, 470*, 55-64.

Pereira, L. S., Cordery, I., & Iacovides, I. (2002). *Coping with water scarcity*: Springer.

Pielke Jr, R. A., & Downton, M. W. (2000). Precipitation and damaging floods: Trends in the United States, 1932-97. *Journal of Climate, 13*(20), 3625-3637.

United Nations Environment Programme (UNEP), & International Environmental Technology Centre (IETC). (1998). *Source book of alternative technologies for freshwater augmentation in some countries in Asia (Technical Publication)* (Vol. no. 8B). Osaka, Japan: UNEP International Environmental Technology Centre.

WWF-UK. (2008). *Summary of the first ever Thames Basin Vulnerability Report.* Retrieved from UK: http://www.wwf.org.uk/wwf_articles.cfm?unewsid=1753

2

Study area: the Mun River Basin, Thailand

2.1 Topography

The Mun River Basin (Figure 2-1), is the largest river basin in Thailand with a total area of 7.1 million ha (71,060 km²). Located in the northeast of the country, it covers 10 provinces: Nakhon Ratchasima (Na), Buri Ram (Bu), Khon Kaen (Kh), Maha Sarakham (Ma), Surin (Su), Roi Et (Ro), Si Sa Ket (Si), Yasothon (Ya), Amnat Charoen (Am) and Ubon Ratchathani (Ub). The basin is bounded on the west and the south by mountain ranges, which are the headwaters of the Mun River and its tributaries. The Mun River meanders and branches significantly (Figure 2-1b), particularly upstream, with 18 primary branches along the main channel. Its cross-section varies between 50 and 480 m. The 726 km river flows east and converges with the Chi River at a district of Ubon Ratchathani Province before reaching its confluence with the Mekong River (Figure 2-1c).

The slopes of the terrain gradually decrease downwards to the Mun River, forming plains for agriculture purpose. On the southern side, hilly areas, with plains in-between them, lie from west to east with declining gradient, following the flow of the river. There are five main landscapes consisting of river levees, flood plains, non-flood plains, undulating land, and hilly areas. The river levees, comparatively small area, are planted with a variety of crops including bamboo, fruit trees, and vegetables and dry-season field crops that are all benefit from receding water. Flood plains, annually inundated, primarily used for rice cultivation, whereas the non-floodplains are planted with rice in the rainy season, and if capillary soil water is available, with a second crop of peanut, soybean, watermelon, etc. after the wet season. Cultivation in the undulating land consists of field crops such as cassava, sugarcane, watermelon, etc. in the upland parts, and wet season rice in the lowland. Lastly, the hilly areas accommodate mostly field crops and fruit trees (Floch, Molle, & Loiskandl, 2007; Limpinuntana, 2001).

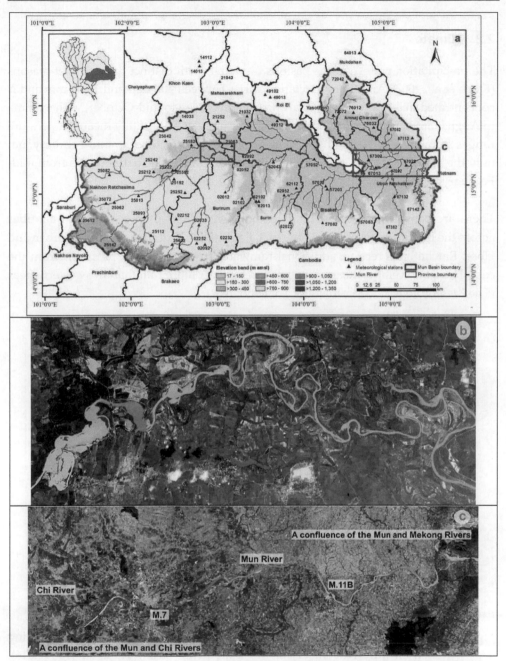

Figure 2-1 The Mun River Basin, Thailand: figure **a** presents its topography, streamflow, 61 selected meteorological stations (53 precipitation stations inside the basin and 8 additional stations from the neighboring stations for supplementary temperature data); figure **b** presents an example of geometry of the river system; figure **c** presents the confluences of the Mun and Chi Rivers, and the Mun and Mekong Rivers.

2.2 Climate

The precipitation pattern in the basin is bi-modal, with distinct dry and wet seasons (Floch et al., 2007). Annual precipitation varies between 800 mm and 1,800 mm and is concentrated in the rainy season starting from mid-May to mid-October, with maxima in August or September (200 mm per month or more). The monthly mean temperature ranges from 25 °C to 30 °C. In the cool season, i.e., mid-October to mid-February, not only the temperature but also the precipitation is lowest. The hot period is from mid-February to mid-May, with the highest temperature in April.

There are considerable spatial and temporal variations of precipitation quantities across the basin. The provinces in the east, i.e., Sisaket, Yasothon, Amnat Charoen and Ubon Ratchathani, receive additional rainwater from tropical depressions from the east. Thus, on average, the eastern provinces collect more rain than the western areas (The Meteorological Department of Thailand, n.d.).

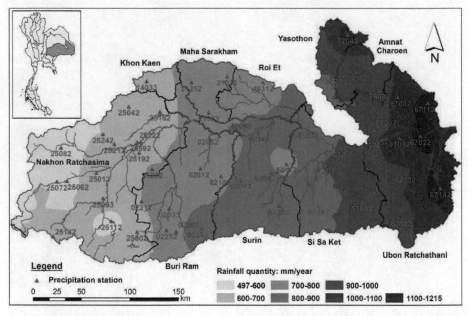

Figure 2-2 Spatial distribution of average total precipitation during the rice-growing season (July-November) at the Mun River Basin, Thailand (1984-2016), created from data from 53 precipitation stations (triangle) selected for the study.

Of the total 196 precipitation stations, only 53 have continuous time-series records of precipitation from 1984-2016. The selected stations are distributed over the basin. The total precipitation during the rice-growing season (July-November) varies considerably. The year-to-year variations have become much larger in the last two decades. The average total precipitation during the rice-growing season ranges from 500-1200 mm (Figure 2-2).

2.3 Rice cultivation

Rice is cultivated on flood plains, non-flood plains, and lowland of undulating areas, occupying 38,565 km^2, which is approximately 75% of the agricultural land, 55% of the basin's total area (Figure 2-3) and about 90% of that is rain-fed (Table 2-1). Only 8% of the total rice land in the basin is irrigated, dispersed over the basin. The irrigated areas are along both sides of the Mun River and its tributaries because all irrigation projects rely on surface water (Figure 2-4). The irrigated land presented here does not include the areas supplied by groundwater because no data concerning irrigated land supplied by groundwater is available.

The KDML105 (Khao Dok Mali 105) and RD6 (Rice Department 6) are the two major varieties of Jasmine rice being grown here with a potential yield of 2.3 t/ha and 4.2 t/ha respectively. They are both medium-maturing types with a life cycle of 120 – 140 days, roughly from July to November (Bureau of Rice Research and Development (BRRD), n.d.). The three growth phases are: vegetative (July-September) from sowing to panicle initiation, reproductive (October) from panicle initiation to flowering and ripening (November) from flowering to full maturity (Brouwer, Prins, & Heibloem, 1989). The common technique for rice cultivation in Thailand nowadays is direct seeding.

The average rice yield in the north-eastern region (2.27 ton/ha) is the lowest in the country (the national average yield is 3.1 ton/ha (FAO, 2018)). In the Mun River Basin, rainfed rice yields are generally below potential due to water shortages. The average rice yields by district vary from 1.84 - 2.99 ton/ha. Higher productivity (> 2.33 ton/ha) was mostly found in the districts in the middle, while most of the districts on the sides of the basin had lower average yields (< 2.33 ton/ha) (Figure 2-5).

Table 2-1 Irrigated rice fields by province in the Mun River Basin.

Province	Rice area (km^2)	Irrigated area (km^2)	% Irrigated area
Nakhon Ratchasima (Na)	6,479	1,260	19.4%
Buriram (Bu)	6,032	472	7.8%
Khon Kaen (Kh)	749	4	0.5%
Maha Sarakham (Ma)	1,872	94	5.0%
Surin (Su)	5,933	298	5.0%
Roi Et (Ro)	2,517	174	6.9%
Si Sa Ket (Si)	5,437	271	5.0%

Province	Rice area (km²)	Irrigated area (km²)	% Irrigated area
Yasothon (Ya)	1,401	69	4.9%
Amnat Charoen (Am)	1,659	79	4.8%
Ubon Ratchathani (Ub)	6,482	381	5.9%
Total	38,561	3,102	8.0%

Provinces are listed from West to East.

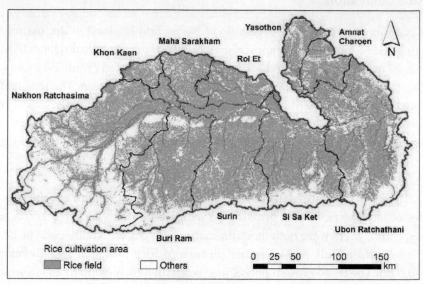

Source: Derived from the land use map obtained from the Land Department, Thailand.

Figure 2-3 Rice cultivation area of the year 2013.

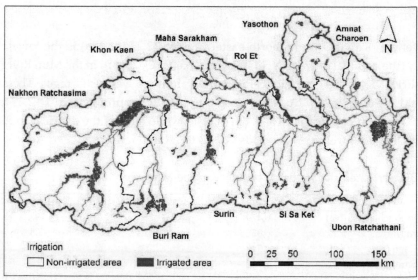

Source: The Royal Irrigation Department, Thailand

Figure 2-4 Irrigation area of the year 2012-2013.

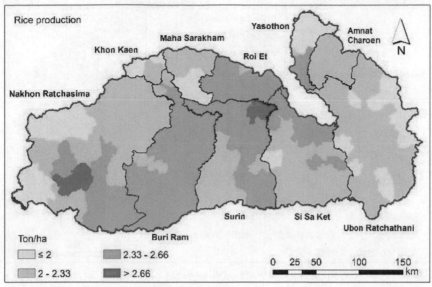

Source: Office of Agricultural Economics, Thailand.

Figure 2-5 Rice production (Ton/ha) by district of the year 2016.

2.4 Available water content in the soil root zone

The available water content of soil represents the amount of water stored in the soil that is available to plants. It is constant for a specific soil type but varies widely between soil texture and structure (Brouwer, Goffeau, & Heibloem, 1985; Wilhelmi & Wilhite, 2002). In the Mun River Basin, soils range from clay, loam and silt loam, which are generally fine-grained soils with high water-holding capacity, to sand and loamy sand, which have coarser textures with low water-holding capacity. Sandy loam is the most common soil in the basin.

The available water content in the soil root zone is a measure of the available water at a specific rooting depth. It is used to identify the ability of soils to buffer crops during periods of low soil moisture (Wilhelmi & Wilhite, 2002). In this case, the effective rooting depth of rice (in the absence of characteristics that can restrict rooting depth, such as bedrock, lithologic discontinuities, water tables) is 1 m (Allen, Pereira, Raes, & Smith, 1998). Based on that, the available water-holding capacities in the soil root zone for rice at the Mun River Basin is presented in Figure 2-6.

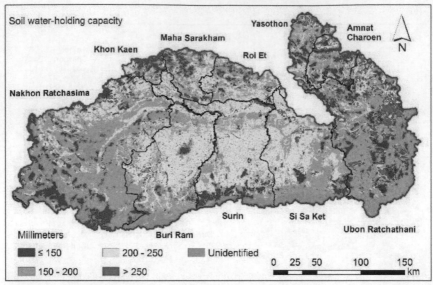

Source: Land Development Department, Thailand.

Figure 2-6 Available water-holding capacity in the soil root zone of the year 2009-
2010.

2.5 Saline soil

A high salt content increases the vulnerability of crops to drought because less water
will be available for plants. Salt reduces the rate and amount of water absorbed by the
plant root. Also, high concentrations of salts are toxic to plants and hinder the
absorption of essential nutrients (Abrol, Yadav, & Massoud, 1988; Brouwer et al.,
1985). Thus, in saline conditions, additional water is needed beyond the crop water
requirement to improve the soil quality.

Significant amounts of saline soil are found in Northeast Thailand. The high, moderate,
and slight salt-affected lands contain salt crusts on the surface at about 10-50%, 1-10%,
and less than 1%, respectively. The total salt-affected area of the basin is approximately
17%, with the lowlands around the Mun River and its tributaries the most affected.
The potential salt-affected area does not have any salt patches on the top, but it has a
rock salt formation underneath. If the thick layer of topsoil is removed or made
thinner, this area will also be salt-affected (Wongsomsak, 1986). This part covers 43%
of the basin, contiguous to the salt-affected land, while the surrounding mountain
ranges are not salt-affected (Wada, 1998) (Figure 2-7).

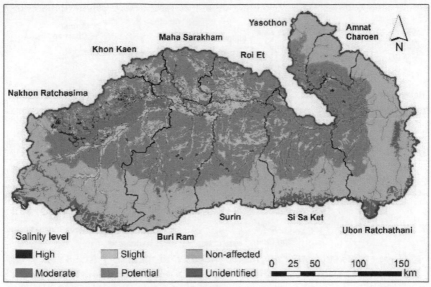

Source: Land Development Department, Thailand.

Figure 2-7 Soil salinity of the year 2006.

2.6 Groundwater

Groundwater is another potential source of irrigation to compensate for shortages of precipitation and surface water; sufficient groundwater can reduce the risk of crop failure during droughts. Besides the amount of groundwater, its quality is also significant because it directly impacts soils, crop growth, and yields. If saline water is applied, salt is left in the soil, which means that more water is required to leach salt from the root zone. Therefore, farmers are less susceptible to water scarcity in areas with abundant groundwater and better quality, whereas in those with less groundwater and poor quality, farmers are more affected during a dry spell.

The groundwater quantity and quality are expressed in terms of expected well yield (m^3/hr.) and total dissolved solids, TDS (mg/l). TDS usually refers to a salinity hazard (Bauder, Waskom, Davis, & Sutherland, 2011; Fipps, 2003). Groundwater with TDS $\geq$ 2000 mg/l is not suitable for irrigation because it will cause reduced yields (Ayers & Westcot, 1985). About 65% of the total area has groundwater with TDS < 500 and 13% with TDS 500-1500 mg/l, meaning it is suitable for irrigation. However, the remaining 22% has TDS > 1500 mg/l, with some values reaching 25,000 mg/l (Department of Groundwater Resources of Thailand, 2015), which is too salty for irrigation (Figure 2-8).

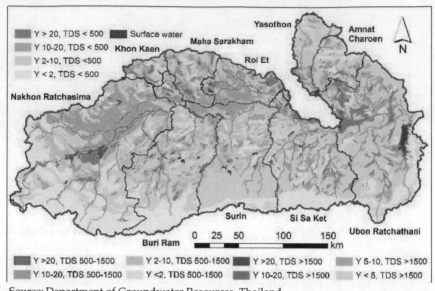

Source: Department of Groundwater Resources, Thailand.

Figure 2-8 Expected well yield (m3/hr.) and TDS (mg/l) of the year 2011-2012.

2.7 Population density

High population density indicates that there are more mouths to feed, and it leads to other dynamic conditions, such as urbanization and less arable land, and thus the need for improved crop productivity. These issues become more problematic in areas with limited resource potential.

The population data were collected at the district level and were averaged on the assumption that people are evenly distributed over the district. Higher population densities were found in the main districts and some districts of Nakhon Ratchasima, Buri Ram, Surin, Si Sa Ket, and Ubon Ratchathani provinces (Figure 2-9).

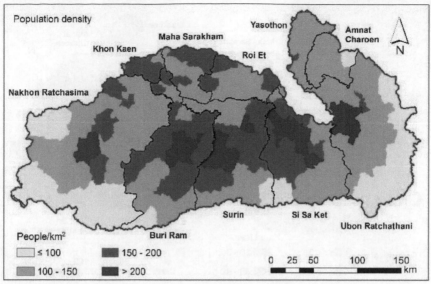

Source: Department of Provincial Administration, Thailand.

Figure 2-9 Population density by district of the year 2016.

2.8 Poverty

Poor people are more vulnerable to disaster risk than others because they lack access to resources and income. They are less able to cope with and adapt to the calamity. Thus, poverty is an important factor of vulnerability (Yodmani, 2001), and it will exacerbate or affect vulnerability levels regardless of the type of hazard (Cardona et al., 2012).

The poverty rate is the proportion of the people whose income falls below the poverty line for the total population of the province. The data distinguish between rural and urban areas, but no digital map providing boundaries between those areas is available. Accordingly, because more than 70% of the total population in all provinces live in rural areas, especially farmers, the rural poverty rate was selected as a proxy for the whole district. The map in Figure 2-10 shows less than 10% of the total population in Khon Kaen, Maha Sarakham, Roi Et and Surin live below the poverty line, whereas some districts in Buri Ram, Si Sa Ket, and Ubon Ratchathani have poverty rates of 30-40%.

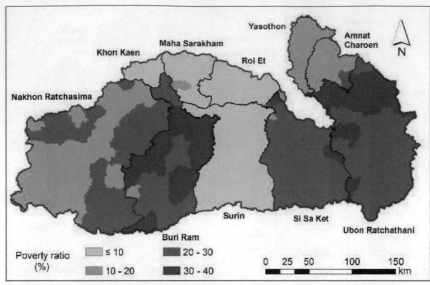

Source: National Statistical Office, Thailand.

Figure 2-10   Poverty rate by district of the year 2013.

2.9 People depending on agriculture

The greater the number of people depending on farming, the higher the vulnerability to droughts will be. This figure includes not only farm laborers (aged 15-64 years), but everyone who depends on agricultural production and income (e.g., children and elderly people).

The percentage of people depending on agriculture reveals the number of people in the province who depend on rice cultivation. In 2016, Amnat Charoen, Maha Sarakham, Roi Et and Yasothon had the largest proportion of people depending on agriculture, whereas Nakhon Ratchasima had the lowest (Figure 2-11).

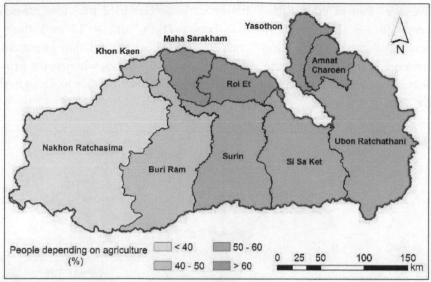

Source: Office of Agricultural Economics, Thailand.

Figure 2-11   People depending on agriculture by province of the year 2016.

2.10 Education level

Educational attainment is a fundamental determinant of human vulnerability to natural hazards. Better-educated people are better able to cooperate and collaborate with experts in designing ways of dealing with and mitigating disaster risk, and they can respond quicker and better to warnings and other public service announcements (Pelling et al., 2004). Moreover, a population with high level of education can easier access to the information source of hazard preparedness, response, reduction, adaptation, and other relevant knowledge, which are increasingly available through new technologies, particularly the internet, nowadays. These motivated and well-informed people can consequently lead to disaster risk reduction (Cardona et al., 2012).

In this study, the average years of schooling of people at the age between 15-59 years, which represents the labor force of the country, were used. The data present the average academic year of all people in the specific provinces; however, they can represent the educational level of farmers because, on average, more than 50% of population in the Mun River Basin rely on agriculture. In the basin, only Khon Kaen province has the average level of education higher than the current Thai national compulsory education of 9 years (Office of the Education Council, 2014), while the other provinces are lack of education.

The average years of educational attainment were classified into four classes: ≤ 6 (Elementary), 6-9 (Junior High), 9-12 (Senior High) and > 12 and they were subsequently scored as 4, 3, 2 and 1, respectively. This indicates that provinces with high average year of schooling are less susceptible to droughts, whereas the provinces with lower average year of schooling are relatively more vulnerable to drought events (Figure 2-12).

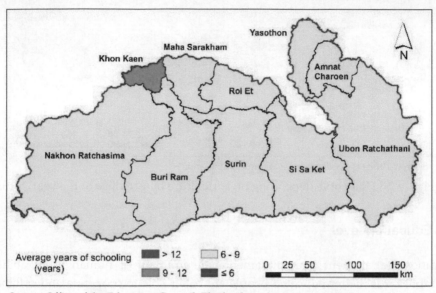

Source: Office of the Education Council, Thailand.

Figure 2-12   Education level by province of the year 2016.

References

Abrol, I., Yadav, J. S. P., & Massoud, F. (1988). *Salt-affected soils and their management*: Food & Agriculture Org.

Allen, R. G., Pereira, L. S., Raes, D., & Smith, M. (1998). FAO Irrigation and Drainage Paper No. 56, Crop Evapotranspiration (Guidelines for Computing Crop Water Requirements). FAO. *Water Resources, Development and Management Service, Rome, Italy*.

Ayers, R., & Westcot, D. (1985). Water quality for agriculture. FAO Irrigation and drainage paper 29 Rev. 1. *Food and Agricultural Organization. Rome*.

Bauder, T. A., Waskom, R. M., Davis, J. G., & Sutherland, P. L. (2011). *Irrigation water quality criteria*: Colorado State University Extension Fort Collins, CO.

Brouwer, C., Goffeau, A., & Heibloem, M. (1985). Irrigation Water Management: Training Manual No. 1-Introduction to Irrigation. *Food and Agriculture Organization of the United Nations, Rome, Italy.*

Brouwer, C., Prins, K., & Heibloem, M. (1989). Irrigation Water Management: Irrigation Scheduling. In *Irrigation water management: Training manual no. 4.* Rome, Italy: Food and Agriculture Organization of the United Nations (FAO).

Bureau of Rice Research and Development (BRRD). (n.d., 14 July 2016). Rice Knowledge Bank (in Thai). *Ministry of Agriculture and Cooperatives, Thailand.* Retrieved from http://www.brrd.in.th/rkb/

Cardona, O. D., van Aalst, M. K., Birkmann, J., Fordham, M., McGregor, G., Perez, R., . . . Sinh, B. T. (2012). Determinants of risk: exposure and vulnerability. In C. B. Field, V. Barros, T. F. Stocker, D. Qin, D. J. Dokken, K. L. Ebi, M. D. Mastrandrea, K. J. Mach, G. K. Plattner, S. K. Allen, M. Tignor, & P. M. Midgley (Eds.), *Managing the Risks of Extreme Events and Disasters to Advance Climate Change Adaptation. A Special Report of Working Groups I and II of the Intergovernmental Panel on Climate Change (IPCC)* (pp. 65-108). Cambridge, UK, and New York, NY, USA: Cambridge University Press.

Department of Groundwater Resources of Thailand. (2015). *Groundwater Situation Report of Thailand.* Retrieved from Bangkok, Thailand:

FAO. (2018). FAOSTAT statistics database. from FAO http://faostat3.fao.org/home/E

Fipps, G. (2003). Irrigation water quality standards and salinity management strategies. *Texas FARMER Collection.*

Floch, P., Molle, F., & Loiskandl, W. (2007). Marshalling water resources: a chronology of irrigation development in the Chi-Mun River Basin, Northeast Thailand. *Colombo, Sri Lanka: CGIAR Challenge Program on Water and Food.*

Limpinuntana, V. (2001). *Physical factors as related to agricultural potential and limitations in northeast Thailand.* Paper presented at the Natural resource management issues in the Korat basin of Northeast Thailand: An overview, Los Banos, Philippines.

Office of the Education Council. (2014). *Map of Provincial Development of Thailand: Education (in Thai)* (Vol. 1). Bangkok, Thailand: Office of the Education Council, Ministry of Education.

Pelling, M., Maskrey, A., Ruiz, P., Hall, P., Peduzzi, P., Dao, Q.-H., . . . Kluser, S. (2004). *Reducing disaster risk: a challenge for development.* Retrieved from New York: United Nations:

The Meteorological Department of Thailand. (n.d.). Meteorological Knowledge (in Thai). Retrieved from https://www.tmd.go.th/info/info.php?FileID=26

Wada, H. (1998). Techniques and strategies to ameliorate salt-affected lands in Northeast Thailand. *JARQ. Japan agricultural research quarterly, 32*(2), 79-85.

Wilhelmi, O. V., & Wilhite, D. A. (2002). Assessing vulnerability to agricultural drought: a Nebraska case study. *Natural Hazards, 25*(1), 37-58.

Wongsomsak, S. (1986). Salinization in Northeast Thailand (< Special Issue> Problem Soils in Southeast Asia).

Yodmani, S. (2001). *Disaster risk management and vulnerability reduction: Protecting the poor*: The Center.

3

Climate variability, trends, and their impacts on rice yields

This chapter is published as:

Prabnakorn, S., Maskey, S., Suryadi, F., & de Fraiture, C. (2018). Rice yield in response to climate trends and drought index in the Mun River Basin, Thailand. *Science of the Total Environment*, 621, 108-119. doi: 10.1016/j.scitotenv.2017.11.136.

Abstract:

Rice yields in Thailand are among the lowest in Asia. In northeast Thailand where about 90% of rice cultivation is rain-fed, climate variability and change affect rice yields. Understanding climate characteristics and their impacts on the rice yield is important for establishing proper adaptation and mitigation measures to enhance productivity. In this paper, we investigate climatic conditions of the past 30 years (1984-2013) and assess the impacts of the recent climate trends on rice yields in the Mun River Basin in northeast Thailand. We also analyze the relationship between rice yield and a drought indicator (Standardized Precipitation and Evapotranspiration Index, SPEI), and the impact of SPEI trends on the yield. Our results indicate that the total yield losses due to past climate trends are rather low, in the range of less than 50 kg/ha per decade (3% of actual average yields). In general, increasing trends in minimum and maximum temperatures lead to modest yield losses. In contrast, precipitation and SPEI-1, i.e., SPEI based on one monthly data, show positive correlations with yields in all months, except in the wettest month (September). If increasing trends of temperatures during the growing season persist, a likely climate change scenario, there is high possibility that the yield losses will become more serious in future. In this paper, we show that the drought index SPEI-1 detects soil moisture deficiency and crop stress in rice better than precipitation or precipitation-based indicators. Further, our results emphasize the importance of spatial and temporal resolutions in detecting climate trends and impacts on yields.

Keywords: Climate variability, Agriculture, Rice yield, SPEI, Mun River Basin, Thailand

3.1 Introduction

Temperature and precipitation are the two fundamental variables commonly used as indicators for changes in climate. The impacts of climate variability and change on crop yields have been studies by numerous researchers worldwide; both for historical and future climates and for various crops (Adams, Hurd, Lenhart, & Leary, 1998; Babel, Agarwal, Swain, & Herath, 2011; Bhatt, Maskey, Babel, Uhlenbrook, & Prasad, 2014; Challinor et al., 2014 etc.; Erda et al., 2005; Fischer, Shah, Tubiello, & Van Velhuizen, 2005; Fuhrer et al., 2006; Hekstra, 1986; Nicholls, 1997). Studies on the impacts of past climate trends on crop yields using empirical models have come to different conclusions depending on crop types and their locations. For example, some studies reported reductions in wheat yields in Russia and France, and maize yields in China as a result of increased temperature (Brisson et al., 2010; Lobell, Schlenker, & Costa-Roberts, 2011; Tao, Yokozawa, Liu, & Zhang, 2008; Wei, Cherry, Glomrød, & Zhang, 2014). Others report increases in wheat yields in Mexico due to decreased nighttime temperature (Lobell et al., 2005), whereas an increase in minimum temperature is the dominant factor attributed to increases Australian wheat yields (Nicholls, 1997). Furthermore, rice yields in China have increased due to significant warming trend (Tao et al., 2008). In the United States, the yield impacts on wheat, maize and soybean are not obvious because of less significant climate trends (Lobell et al., 2011). In a study in the Koshi basin (Nepal) Bhatt et al. (2014) pointed out that crop yield impacts differ even in the same basin depending on altitudes.

These findings have resulted in an improved understanding of the links between climate and crop yields and the extent to which climate impacts productivity. However, these studies used basic climatic variables such as minimum, maximum and mean temperatures and precipitation at a rather coarse temporal (annual or by growing season) and spatial scale, such as global (Lobell & Field, 2007; Lobell et al., 2011), regional (Lobell, Cahill, & Field, 2007; Schlenker & Lobell, 2010), and national scale (Nicholls, 1997; Rowhani, Lobell, Linderman, & Ramankutty, 2011; Wei et al., 2014).

Drought indices relate to cumulative effects of a prolonged and abnormal moisture deficiency (World Meteorological Organization, 1992), thus they have a strong connection to agriculture. There are a number of drought indices commonly used such as Standardized Precipitation Index (SPI) (McKee, Doesken, & Kleist, 1993), Standardized Precipitation and Evapotranspiration Index (SPEI) (Vicente-Serrano, Beguería, & López-Moreno, 2010), Palmer Drought Severity Index (PDSI) (Palmer, 1965), Normalized Difference Vegetation Index (NDVI) (Tucker, 1979).

The SPEI, a meteorological drought index, is more relevant to agriculture than precipitation or precipitation-based indices such as SPI, because it is based on both precipitation and temperature. It can be computed at any preferable timescales and the values represent both wet and dry conditions (Guttman, 1999; Zargar, Sadiq, Naser, & Khan, 2011), which both can affect crop yields when threshold values are exceeded. In addition, the meteorological drought index can detect the onset of drought sooner than agricultural and hydrological drought indices. However, even though SPEI has been widely used for monitoring and forecasting climate variations and conditions, it is rarely used to evaluate the link between crop yields and climate.

The objectives of this paper are to examine climate variability and trends, and the relationships between rice yields and basic climatic parameters such as minimum (Tmin), average (Tave), maximum (Tmax) temperatures, precipitation (Prec), and the drought index SPEI. This study is implemented for the Mun River Basin, Thailand, where no such study has been executed before. The study advances previous work by assessing yield changes at monthly time step rather than using averages or summations over the growing season. A shorter time step is important because of the varying degree of crop sensitivity to climate at each growth stage. In addition, this study takes a finer spatial scale than earlier studies, using basin and sub-basin scales as opposed to national level and global assessments. Because the climate - yield relationship is scale dependent and empirical models at global scale cannot reliably be used to anticipate the outcomes at finer scales (Lobell & Field, 2007; Tao et al., 2008), assessments on a smaller spatial scale (such as basin level) deepens the understanding of crop-climate links.

3.2 Data and methods

3.2.1 Data collection

Time series data of monthly precipitation, Tmin and Tmax from 1984 to 2013 were obtained from the Royal Irrigation Department and the Meteorological Department of Thailand. Of the total 196 precipitation stations in the basin, only 53 have continuous time-series records for the specified period. Among them, only 10 stations also have adequate temperature data. Therefore, data from several adjacent stations outside the basin were included. Annual rice production data of the 10 provinces were acquired from the Office of Agricultural Economics, Ministry of Agriculture and Cooperatives. The length of the yield records corresponds with the length of climatic records, except for the Amnat Charoen province where the recording of data only started in 1994.

3.2.2 Drought index calculation and classification

The SPEI is based on monthly differences between precipitation and potential evapotranspiration (PET). There are different methods to compute PET varying from simple methods, such as Thornthwaite (1948) or Hargreaves (1994), to sophisticate ones, such as Penman-Monteith (PM) (Allen, Pereira, Raes, & Smith, 1998). Though the PM method has been accepted as the standard method by the Food and Agriculture Organization of the United Nations (FAO), the International Commission for Irrigation and Drainage (ICID) and the World Meteorological Organization (WMO), here we choose the Hargreaves method because of data limitations. Moreover, Mavromatis (2007) concluded that for the computation of drought indices, simple PET estimation methods provide outcomes similar to the complex methods. Santiago Beguería, Vicente-Serrano, Reig, and Latorre (2014) compared the SPEI values calculated with three different PET estimation methods: Penman-Manteith, Hargreaves, and Thornthwaite. They found that the differences are small in humid regions and recommended the Hargreaves equation over the Thornthwaite equation in data scarce areas.

Water deficit or surplus for a specific month i is calculate using:

$$D_i = P_i - PET_i,$$ (3.1)

where Di is the water deficit or surplus at month i (mm/month), and Pi is effective precipitation (mm/month).

The D_i values were aggregated at different time scales, following the same procedure as that for the SPI. The log-logistic distribution was used to standardize the D series to derive the SPEI values at preferred time scales. The SPEI at 1- and 3-month timescales were considered because they are relevant to agriculture (Potop, Boroneanţ, Možný, Štěpánek, & Skalák, 2014). Monthly SPEIs of each province over 30 years were determined, but only the months of the rice-growing season (July to November) were used in further analysis.

Following previous studies (Santiago Beguería et al., 2014; Dorman, Perevolotsky, Sarris, & Svoray, 2015; Wang, Zhu, Xu, & Liu, 2015), the R package was used to compute PET and SPEI, developed by S Beguería and Vicente-Serrano (2013). The classification of droughts using the SPEI values given in Table 3 follows the same criteria as the SPI due to their similarity in fundamental principles and calculation (Tan, Yang, & Li, 2015).

Table 3-1 Dry/wet conditions corresponding with the seven SPEI categories

SPEI value	Category
≥ 2.00	Extreme wet (EW)
1.50 to 1.99	Severe wet (SW)
1.49 to 1.00	Moderate wet (MW)
0.99 to -0.99	Normal (N)
-1.00 to -1.49	Moderate drought (MD)
-1.50 to -1.99	Severe drought (SD)
≤ -2.00	Extreme drought (ED)

Source: Potop et al. (2014)

3.2.3 Correlation, relationship and impact analysis

Correlations of rice yield with climatic variables were estimated following an established approach as described in (Bhatt et al., 2014; Lobell & Field, 2007; Lobell et al., 2005; Nicholls, 1997). This approach removes confounding effects of long-term variations in yields, such as cultivars, crop management and fertilizers, by calculating the first differences in the yield, climatic variables and drought index (VAR_t – VAR_{t-1}). In the next step, the Pearson product-moment correlation coefficients were calculated. The correlation results provide initial information on the positive or negative sign of relationships which helps understanding the regression results.

Subsequently, the relationships of rice yield with the climate variables and with the SPEIs were investigated using a multiple linear regression model. Though less complex than crop simulation models, the multiple linear regression model is able to capture the net climate effects of combined climate variables at monthly time steps during the growing season (Lobell & Field, 2007). The intercept was forced through zero to avoid trend effects (Nicholls, 1997). The multivariate linear regression is of the form:

$$\Delta Y_i = \gamma_{1i}*\Delta Tmin + \gamma_{2i}*\Delta Tmax + \gamma_{3i}*\Delta Prec, \qquad (3.2)$$

where ΔY_i is first differences in annual rice yield of province i (t/ha), $\Delta Tmin$ and $\Delta Tmax$ are first differences in minimum and maximum temperatures (°C), $\Delta Prec$ is first differences in precipitation (mm), and γ is a vector of estimated coefficients. Similarly, the approach was applied to the SPEIs as following:

$$\Delta Y_i = a_{1i}*\Delta SPEI \qquad (3.3)$$

where $\Delta SPEI$ is first differences in SPEI values and a is a vector of estimated coefficient.

Bootstrapping technique (Fox, 2015; Lobell & Field, 2007) was employed to overcome size limitation of the 30 years timeseries data by resampling 1000 bootstrap samples (independent random sample). By re-calibrating estimated coefficients, this technique provides more precise statistical inferences of the coefficients, including their relevant 90% confident intervals.

After running the regression and determining the regression coefficients, the first differences values were substituted by the trend slopes of climatic variables and of SPEI. This provide the impact of climate trends on crop yields. Because climatic trends affect crop development in different growth stages differently, the analysis was carried out not only for the average values over rice-growing season, but also for each individual month of the growing period.

3.3 Results

3.3.1 Observed temporal variations of wet and dry conditions

The analysis of the SPEI indicates that over the past 30 years the basin encountered both extreme wet and extreme dry conditions (Figure 3-1). The dry/wet conditions as indicated by the 1-month SPEI (SPEI-1) changed more frequently than shown by the 3-month timescales because the SPEI-1 values represent the conditions of each individual month, and thus the influence of dry/wet conditions does not affect the next consecutive month. On the other hand, in the computation of the SPEI-3, extreme weather conditions of one month are visible in SPEI values of the next two consecutive months. This makes the duration of extreme events as indicated by the SPEI-3 longer than as shown by SPEI-1.

During the months of the rice-growing season, all dry and wet conditions occurred (Table 3-2). Approximately 35% of the total number of months experienced anomalous dry or wet conditions. The proportion of wet months was slightly higher than the dry ones. The number of extreme wet and dry months increased considerably when the time scale increased from 1 month to 3 months (i.e. SPEI-1 to SPEI-3). This shows that extreme events may not always be detected effectively with the 1-month SPEI because impacts accumulate and become more visible over time. Therefore, the SPEI-1 in combination with SPEI-3 provide a more complete description of dry/wet conditions.

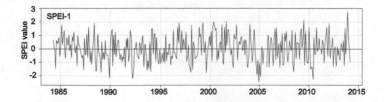

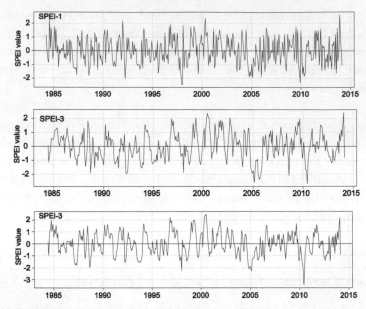

Figure 3-1 Examples of temporal variations of SPEI values at 1- and 3- month
timescales of Si Sa Ket province (left) and Ubon Ratchathani province (right)
for 1984-2013

Table 3-2 Number of months during rice-growing period at different SPEI
classifications of all ten provinces at 1- and 3- month timescales.

Category	SPEI-1	SPEI-3
Extreme Wet (EW)	18	26
Severe Wet (SW)	91	90
Moderate Wet (MW)	157	144
Normal (N)	985	983
Moderate Dry (MD)	167	166
Severe Dry (SD)	73	65
Extreme Dry (ES)	9	26
Total Wet	266	260
Total Dry	249	257

Table 3-3 Spatial and temporal variations of climatic characteristics of rice-growing months that are classified based on SPEI-1 values.

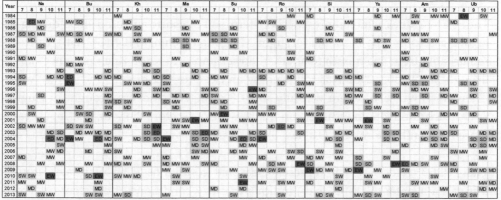

The provinces are presented from west to east according to their geographical locations. The numbers at the top of the table indicate the months (July to November). The climatic events are classified into two main groups: the wetness (blue) and dryness (orange). The colour levels display the degree of conditions ranging from moderate to extreme.

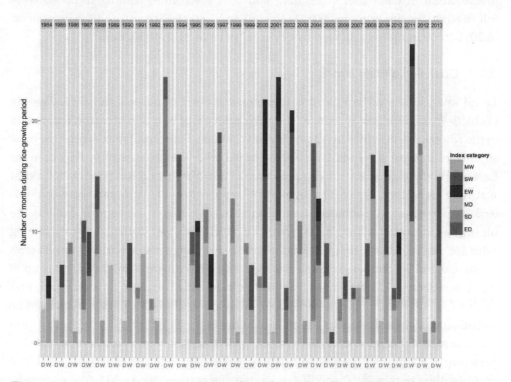

Figure 3-2 Annual variations in the number of months affected by dry/wet conditions (SPEI ≤ -1 and SPEI ≥ 1) summed for all provinces from 1984-2013. The orange and blue colours present dry and wet conditions, respectively.

Different shades of the colours define the intensities of dry (D) and wet (W) states. The classification is based on the SPEI-3 values

The SPEI-1 values, which represent short-term soil moisture and crop stress during the growing season (The National Drought Mitigation Center (NDMC), n.d.), fluctuate remarkably from one month to another. For example, in Khon Kaen province, severe drought in October 2002 was followed by an extreme wet month, and in the Amnat Charoen province, a severe wet July in 2013 changed to severe drought in August. Such quick variations in weather conditions were rare before the year 2000, but became more frequently afterwards (Table 3-3).

The SPEI-3 values, which represent seasonal conditions (The National Drought Mitigation Center (NDMC), n.d.) indicate that dry conditions of different categories were more prominent in the period 1986–1999, whereas wet conditions were more prominent in the period after 1999 (Figure 3-2). Another interesting observation is the alternate occurrences of floods and droughts in recent years. For example, high precipitation in years 2001 (1380 mm) and 2002 (1400 mm) leading to floods were followed by the dry years in 2003 (1140 mm) and 2004 (1090 mm), and the great flood in 2011 (1460 mm) was followed by dry conditions in 2012 (1030 mm).

3.3.2 Observed climate trends

Trend analysis shows that climate variables and SPEI trends are mostly increasing (Table 3-4), in particular, those with a significant level of 90% or better (the Tmin trend in the Khon Kaen province in August is an exception). Tmin and Tmax exhibit the most significant increasing trends as compared to the other variables, with provinces in the East exhibiting clearer warming trends than in the West. Tmin has a more significant warming trend than Tmax. The significant trend values of Tmin vary between -0.1-0.65 °C per decade, distributed over all months. The month November exhibits the highest warming trend, ranging between 0.45 and 0.65 °C per decade. Lower warming rates are observed in the first three months of the growing season, with the lowest rates in August and September. Tmax shows an even higher upward trend with rates up to 0.8 °C per decade. The statistically significant increasing trends in Tmax occur mostly in October (0.3-0.6 °C per decade) and November (0.45-0.8 °C per decade). The statistically significant trends in precipitation and SPEI-1 are all increasing and concentrated in August and September, with increases in precipitation varying between 10-60 mm per decade and highest values occurring in the Maha Sarakham and Yasothon provinces. The SPEI-1 exhibits an increasing trend of 0.2-0.5 per decade.

Table 3-4   Trends in Tmin, Tmax, precipitation, and 1-month SPEI per decade for 1984-2013.

| | \multicolumn{6}{c}{Tmin} | \multicolumn{6}{c}{Tmax} |
	Jul	Aug	Sep	Oct	Nov	Mean	Jul	Aug	Sep	Oct	Nov	Mean
Na	0.34***	0.30***	0.30***	0.47***	0.64**	0.41***	-0.06	-0.03	-0.13	0.26	0.47'	0.10
Bu	0.22**	0.24***	0.22**	0.23'	0.27	0.24**	0.06	0.13	0.04	0.34*	0.60*	0.23*
Kh	-0.01	-0.10*	0.05	0.13	0.28	0.07	0.06	-0.07	0.01	0.42*	0.50*	0.19'
Ma	-0.03	-0.08	0.09	0.27	0.47'	0.15	0.31*	0.09	0.22'	0.60**	0.78**	0.40**
Su	0.26**	0.20**	0.19**	0.25'	0.38	0.26**	0.12	0.09	0.05	0.38*	0.58*	0.24**
Ro	0.13	0.11*	0.22***	0.23	0.38	0.21**	0.18	0.08	0.06	0.38*	0.53*	0.24**
Si	0.28**	0.22**	0.20**	0.38*	0.47'	0.31***	0.17	0.10	-0.07	0.29*	0.45'	0.19*
Ya	0.28**	0.25***	0.29***	0.34*	0.52'	0.33***	0.39*	0.30**	0.30*	0.58**	0.77**	0.47***
Am	0.28**	0.25***	0.29***	0.34*	0.52'	0.33***	0.39*	0.30**	0.30*	0.58**	0.77**	0.47***
Ub	0.20'	0.16'	0.16'	0.19	0.46*	0.24*	0.28*	0.28**	0.11	0.58**	0.77**	0.41***

| | \multicolumn{6}{c}{Precipitation} | \multicolumn{6}{c}{SPEI1} |
	Jul	Aug	Sep	Oct	Nov	Sum	Jul	Aug	Sep	Oct	Nov	Mean
Na	4.96	15.29'	16.70	9.64	0.53	9.42'	0.18	0.43*	0.36'	0.09	0.06	0.22*
Bu	5.68	19.02	12.10	10.78	1.04	9.73*	0.15	0.35'	0.16	0.13	-0.08	0.14
Kh	21.89	22.96	-5.83	5.35	1.18	6.08	0.23	0.17	-0.07	0.03	-0.20	0.06
Ma	21.33	36.06**	62.23*	-2.04	4.29	24.37**	0.21	0.45*	0.51*	-0.16	-0.08	0.19'
Su	14.88	25.01'	20.10	0.83	1.03	12.37*	0.22	0.36'	0.18	-0.06	-0.10	0.12
Ro	17.30	3.91	44.01'	-1.75	5.38	14.60	0.13	0.04	0.34	-0.06	-0.04	0.09
Si	23.58	2.07	27.18	-1.31	-6.68	8.97	0.22	0.01	0.28	-0.08	-0.16	0.06
Ya	38.53	9.71	65.04*	-3.34	1.52	22.29'	0.27	-0.08	0.50*	-0.14	-0.20	0.07
Am	15.23	-23.78	22.17	-8.46	3.04	1.64	0.18	-0.23	0.20	-0.25	-0.12	-0.04
Ub	25.63	-19.57	25.40	-11.49	-1.06	3.78	0.21	-0.19	0.26	-0.23	-0.14	-0.02

*** p-value ≤ 0.001, ** p-value ≤ 0.01, * p-value ≤ 0.05, ' p-value ≤ 0.1

The red and blue colors represent negative and positive values respectively.

Provinces are listed from west to east.

3.3.3   Climate – rice yield correlations

3.3.3.1   Correlation rice yields with temperatures

Our results show that rice yields are generally negatively correlated with temperatures Tmin, Tmax and Tave (Table 3-5). The correlation of rice yield with Tmax, Tmin, and Tave are comparable and predominantly negative at the significance level 90% or higher (p-value ≤ 0.1), except for Tmin in the Ubon Ratchathani province in November (Table 3-6). Tmax exhibits a slightly stronger correlation with the rice yield than Tmin and Tave, which is in agreement with the findings of Bhatt et al. (2014). The strongest correlations with Tmin are found in November (the harvest month), especially in the eastern provinces, whereas the highest correlations with Tmax could be found in any month.

Table 3-5 Number of months with positive and negative significant correlations at
90% and 95% significant levels.

Correlation	Significant level	Tmin	Tmax	Tave	Rainfall	SPEI1	SPEI3
Positive	≥ 95% (p-value ≤ 0.05)	0	0	0	2	2	3
	≥ 90% (p-value ≤ 0.10)	1	0	0	7	8	7
Negative	≥ 95% (p-value ≤ 0.05)	7	11	10	0	0	0
	≥ 90% (p-value ≤ 0.10)	14	15	15	2	4	0

The mean of Tmin, Tmax and Tave over the growing season exhibit correlations
comparable to those for individual months. The negative correlations in Jul, Oct and
Nov dominate some of the positive correlations observed in Aug and Sep resulting in
negative correlations with the seasonal mean in all provinces except Ub (Table 3-6). All
the positive correlations observed with temperatures are however statistically
insignificant at 90% or higher confidence level, except in Nov with Tmin in Ub
province. The mean of Tmin values over the growing season exhibit significant
correlations with yield in four provinces (Na, Kh, Ro and Ya), out of the total seven
provinces where individual months show significant correlations with the yield (Na,
Kh, Ro, Si, Ya, Am and Ub). For the mean Tmax this occurred in five (Na, Kh, Ro, Si
and Ya) out of eight provinces (Na, Kh, Ma, Ro, Si, Ya, Am and Ub). This shows that
mean temperature values over the growing season provide satisfactory results in the
analysis as well as temperature values of individual months.

Table 3-6 Pearson's correlation coefficients between rice yield and monthly minimum
temperature (Tmin), maximum temperature (Tmax), mean temperature
(Tave), and precipitation for 1984-2013.

	Tmin						Tmax					
	Jul	Aug	Sep	Oct	Nov	Mean	Jul	Aug	Sep	Oct	Nov	Mean
Na	-0.35*	-0.42**	-0.06	-0.09	-0.18	-0.33*	-0.18	-0.53**	-0.41**	-0.27	-0.08	-0.42**
Bu	-0.26	-0.04	-0.09	-0.21	0.01	-0.15	-0.16	0.03	0.13	-0.17	-0.12	-0.15
Kh	-0.14	-0.10	-0.30	-0.34*	-0.30	-0.41**	-0.03	-0.22	0.09	-0.48**	-0.50**	-0.46**
Ma	-0.03	0.11	0.07	-0.18	-0.30	-0.18	-0.31*	-0.11	0.06	-0.25	-0.11	-0.23
Su	-0.01	0.10	0.01	0.03	-0.26	-0.16	-0.06	0.09	0.31	-0.05	0.02	0.07
Ro	-0.33*	0.05	-0.38**	-0.21	-0.27	-0.39**	-0.28	-0.42**	-0.03	-0.29	-0.10	-0.35*
Si	-0.10	0.13	-0.06	0.02	-0.33*	-0.24	-0.19	-0.01	0.01	-0.47**	-0.18	-0.41**
Ya	-0.22	-0.36*	-0.06	-0.07	-0.39**	-0.37**	-0.05	0.00	-0.16	-0.32*	-0.29	-0.41**
Am	-0.21	-0.11	0.22	0.16	-0.51**	-0.27	0.23	0.20	0.23	-0.18	-0.39*	-0.08
Ub	-0.33*	0.05	-0.03	0.07	0.35*	0.14	-0.38**	0.14	-0.20	0.25	0.19	0.05

	Tave						Precipitation					
	Jul	Aug	Sep	Oct	Nov	Mean	Jul	Aug	Sep	Oct	Nov	Sum
Na	-0.28	-0.54**	-0.37*	-0.26	-0.15	-0.44**	0.30*	0.03	0.33*	-0.18	-0.08	0.18
Bu	-0.23	0.01	0.04	-0.25	-0.07	-0.19	0.01	0.11	-0.28	0.26	0.06	0.04
Kh	-0.08	-0.20	-0.03	-0.49**	-0.44**	-0.50**	-0.01	0.19	-0.38*	-0.05	0.21	-0.08

	Tave						Precipitation					
	Jul	Aug	Sep	Oct	Nov	Mean	Jul	Aug	Sep	Oct	Nov	Sum
Ma	-0.22	-0.01	0.08	-0.27	-0.24	-0.25	0.07	0.21	-0.30	0.06	-0.25	-0.09
Su	-0.05	0.11	0.24	-0.02	-0.13	-0.04	0.03	0.37*	-0.33*	0.14	-0.24	-0.02
Ro	-0.32*	-0.33*	-0.18	-0.31*	-0.21	-0.40**	0.35*	0.20	-0.27	0.08	-0.06	0.13
Si	-0.18	0.05	-0.02	-0.29	-0.28	-0.37*	0.02	0.23	-0.06	0.27	-0.26	0.14
Ya	-0.12	-0.12	-0.15	-0.26	-0.36**	-0.43**	-0.08	0.49**	-0.06	0.37**	-0.04	0.31*
Am	0.09	0.12	0.27	-0.04	-0.49**	-0.18	0.30	-0.05	-0.15	0.31	-0.07	0.11
Ub	-0.42**	0.13	-0.18	0.19	0.29	0.11	0.16	-0.08	0.04	-0.13	0.01	0.04

* p-value ≤ 0.10, ** p-value ≤ 0.05
The red and blue colours present the negative and positive values, respectively

3.3.3.2 Correlation rice yields with precipitation

The correlations of rice yields with precipitation are predominantly positive, except in September, but are generally weaker than with temperatures (Table 3-6). The highest correlations can be found in any month, except November. In November (harvest month) the least amount of water is required and water shortage during the last 15 days will not affect yields. On the contrary, dry conditions will homogenize maturation and facilitate harvesting (Wopereis, Defoer, Idinoba, Diack, & Dugué, 2008).

The summation of monthly precipitation over the growing season exhibit only one significant correlation with the yield, although several significant correlations are observed with individual months. Because of large fluctuations between months, average or summation of precipitation over the growing season is not representative. This shows the importance of monthly values as opposed to annual or seasonal aggregations or averages.

3.3.3.3 Correlation rice yield with drought index SPEI

Yields are stronger correlated with the SPEI-1 than with the monthly precipitation and SPEI-3. More months exhibit a significant correlation with SPEI-1 than with precipitation or SPEI-3 (Table 3-5). The correlations with the 1-month SPEI exhibit a similar pattern to those with precipitation but stronger. The rice yield in the provinces on the west, (Na, Bu, Kh, Ma, and Su) clearly correlated with the SPEI-1 in September, while the high correlations of eastern provinces were in most of the months, but not in September and November (Table 3-7).

Table 3-7   Pearson's correlation coefficients between rice yield and 1- and 3-month SPEI for 1984-2013.

	SPEI1						SPEI3					
	Jul	Aug	Sep	Oct	Nov	Mean	Jul	Aug	Sep	Oct	Nov	Mean
Na	0.33*	0.12	0.34*	-0.03	-0.05	0.29	0.23	0.29	0.46**	0.20	0.11	0.35*
Bu	0.03	0.10	-0.32*	0.18	0.14	0.07	0.03	0.12	-0.12	0.01	-0.02	0.01
Kh	0.02	0.21	-0.38*	-0.03	0.25	0.03	-0.15	0.08	-0.05	-0.05	-0.25	-0.12
Ma	0.07	0.24	-0.33*	0.07	-0.23	-0.10	0.07	0.46**	-0.10	-0.14	-0.28	0.01
Su	0.02	0.36*	-0.34*	0.14	-0.20	-0.03	-0.14	-0.02	-0.04	0.00	-0.25	-0.15
Ro	0.35*	0.22	-0.28	0.09	-0.04	0.14	0.32*	0.37*	0.16	0.00	-0.19	0.16
Si	-0.02	0.25	0.00	0.32*	-0.17	0.17	0.38**	0.28	0.10	0.19	0.04	0.28
Ya	-0.04	0.44**	-0.04	0.41**	0.06	0.34*	-0.03	0.33*	0.18	0.29	0.13	0.25
Am	0.28	-0.04	-0.18	0.36	0.02	0.22	0.30	0.17	-0.12	-0.12	-0.03	0.04
Ub	0.18	-0.03	0.08	-0.15	0.04	0.07	0.25	0.25	0.06	-0.06	-0.02	0.15

* p-value ≤ 0.10, ** p-value ≤ 0.05

3.3.4   Climate impacts on rice yield

3.3.4.1   Regression with climatic variables

The regression results show that changes in climatic variables (Tmin, Tmax and precipitation) could explain the variance in year-to-year rice yield changes from 22% to 37% (Table 3-8). The unexplained yield variance reveals the importance of variables disregarded in this analysis, such as changes in financial status or other conditions that affect field management practices, as well as data uncertainty.

The regression results for a particular month gives more explanatory power (R^2) than for the average over the entire growing season. Those particular months may coincide with crucial crop development stages that directly influence yields, such as, tillering from the second half of August to mid-September and panicle differentiation from mid-September to October. Further, regressions with Tmin and Tmax separately gave more satisfactory outcomes (higher R^2) than using the mean temperature (Tave).

Table 3-8 R^2 values of the multiple linear regression models.

	Na	Bu	Kh	Ma	Su	Ro	Si	Ya	Am	Ub
Rice yields with Tmin, Tmax and Prec	0.31**	0.22*	0.37**	0.11	0.26*	0.26**	0.26**	0.33**	0.31*	0.19
Rice yields with Prec	0.11*	0.07	0.14*	0.08	0.13*	0.12*	0.07	0.13*	0.09	0.03
Rice yields with SPEI1	0.12*	0.10*	0.14*	0.11*	0.12*	0.12*	0.10*	0.16*	0.13	0.03

* p-value ≤ 0.10, ** p-value ≤ 0.05

The regression results confirmed that rice yields respond negatively to the increases in Tmin and Tmax. This is in line with the previous findings from the U.S. (Edmonds & Rosenberg, 2005), and at global scale (Lobell & Field, 2007). Rice yields are more susceptible to changes in Tmin than in Tmax. However, in some provinces contradictory effects occur between increases in nighttime (Tmin) and daytime (Tmax) temperatures.

The relationships between yield and SPEI-1 reveal higher R^2 values as compared to yield and precipitation (Table 3-8). In 8 out of 10 provinces, the relationship of yield with SPEI-1 was statistically significant as compared to 5 out of 10 provinces for precipitation. The SPEI-1 can explain the year-to-year variance in rice yield for 10% to 16%. The coefficients of both precipitation and 1-month SPEI have the same direction, mostly positive, except in September.

3.3.4.2 Impact of climate trends on rice yields

Overall, observed climate trends suppress rice yields in all provinces, except in the Su province where the yield increased by 70 kg/ha per decade (Figure 3-3). The total yield losses of each province are mostly below 50 kg/ha ($\approx$ 3% compared to the average yield) per decade, driven by temperature trends rather than precipitation trends. In most cases, yields decline due to increasing trends of Tmin and Tmax. The reduction rate by province varied from 2 - 10% for each 1 °C increase in both Tmin and Tmax during the growing season.

The 90% confident intervals, which present uncertainty of the regression models in the estimation of yield impacts, show large ranges, crossing in some instances the yield impact = 0 line. These are due in part to opposing influences between temperatures and precipitation and between Tmin and Tmax as well as data uncertainty. Consequently, in some provinces it is difficult to determine a clear direction of total yield impact due to climate trends. The largest uncertainty occurs in Amnat Charoen province due to limited data availability (Figure 3-3).

Estimated yield impacts due to trends in precipitation and SPEI-1 (Figure 3-4) show clear directions, though varying from one province to another. In the Si Sa Ket, Yasothon, and Amnat Charoen provinces (in the east) rice yields decrease as a result of observed trends in precipitation and in SPEI-1 while in the Ub, Ro, Su and Na provinces yields increase. Yield gains tend to occur in July/August/September whereas losses occur later in the season (September/October).

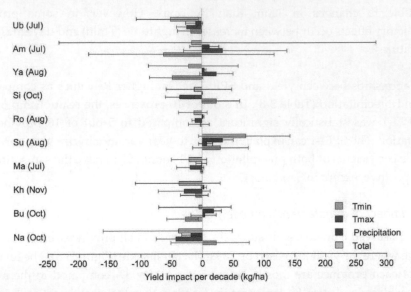

Figure 3-3 Estimated impacts of climate trends from the multiple linear regression for
1984-2013 on rice yield (kg/ha/decade). Only the months with the highest
R^2 for the given provinces are presented. Most other months are statistically
insignificant. The orange, red, and blue bars express the yield impacts due
to trends in Tmin, Tmax and precipitation, respectively. The grey bars
present the total yield impacts due to trends in all climatic variables. The
error bars indicate 90% confident interval.

Another clear result is that yield losses due to trends in SPEI-1 are larger than those
due to precipitation trends. On the other hand, observed yield gains due to trends in
SPEI-1 are usually smaller than those due to trends in precipitation. Therefore, the
SPEI-1 is more responsive to rice yield changes, particularly when yields decrease.

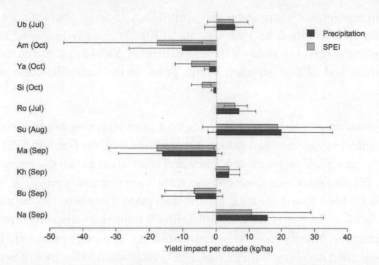

Figure 3-4 Estimated impacts of precipitation and 1-month SPEI trends from simple linear regression of rice yield with precipitation and with 1-month SPEI for 1984-2013 on rice yield (kg/ha/decade). Only the months with the highest R2 for the given provinces are presented, most other months are statistically insignificant. The blue and pink bars indicate the impacts due to trends in precipitation and 1-month SPEI, respectively. The error bars present 90% confident interval.

3.4 Discussion

We show, in agreement with (Jagadish et al., 2010), that assessing yield changes using minimum and maximum temperatures separately provides less uncertainty than using average temperatures. There is evidence that Tmin and Tmax provide different effects on crop phenological development and physiological processes (Wassmann et al., 2009). Our results indicate that rice yields are more vulnerable to changes in Tmin than in Tmax, consistent with Peng et al. (2004). The result corroborates some literature (Nagarajan et al., 2010; Pathak et al., 2003; Welch et al., 2010) that higher Tmin contributes to lower rice yield. However, the physiological mechanism responsible for the negative impacts of Tmin on rice yields remains unclear (Peng et al., 2004; Wassmann et al., 2009). Higher maximum temperatures especially over 35 °C and lasting more than 1 hour during anthesis and flowering stages induces spikelet sterility (Yoshida, Satake, & Mackill, 1981), which finally results in yield reduction.

Our results indicate that increases in minimum and maximum temperatures generally have negative impacts on rice yield. However, in some provinces opposing impacts of nighttime (Tmin) and daytime (Tmax) in the multiple regression models in some growth stages indicate that the impacts of Tmin and Tmax on rice development and yield remain ambiguous. Ultimately, this contributes to large uncertainties in total yield impacts as also found by Welch et al. (2010). Accurately assessing effects of temperature change on crop yields is therefore difficult. Yield impacts due to changes in precipitation and SPEI-1 are clearer and point in the same direction with less uncertainty.

Rice yield are stronger correlated with the SPEI-1 than with precipitation and SPEI-3 (both in statistical significance and value). We conclude that, the 1-month SPEI is better capable of detecting soil moisture deficiency and crop stress in rice than precipitation and SPEI-3. We also examined the correlations between rice yields with 2, 6, 9 and 12-month SPEIs (Table 3-9) and the results show inferior to those between the yields and SPEI-1 and SPEI-3. The correlations with the SPEI-1 follows a similar pattern to those with precipitation (mostly positively correlated, except in September). In some provinces, the yield decreases with increases of precipitation in the month September, even though the monthly precipitation is lower than crop water requirement. This is due to extremes in precipitation intensity, which according to the precipitation records (used in this study) could be as high as 250 mm per day.

Table 3-9 Pearson's correlation coefficients between rice yield and 2, 6, 9 and 12-month SPEI for 1984-2013.

	SPEI2						SPEI6					
	Jul	Aug	Sep	Oct	Nov	Mean	Jul	Aug	Sep	Oct	Nov	Mean
Na	0.24	0.35*	0.32*	0.16	-0.12	0.29	0.14	0.21	0.39**	0.27	0.26	0.30
Bu	0.10	0.08	-0.17	-0.07	0.22	0.07	-0.07	-0.05	-0.17	0.02	0.08	-0.03
Kh	-0.15	0.22	-0.05	-0.34*	0.12	-0.07	-0.40**	-0.13	-0.27	-0.14	-0.09	-0.28
Ma	0.30	0.29	-0.17	-0.26	-0.04	0.05	-0.03	-0.01	-0.21	-0.07	0.04	-0.08
Su	-0.22	0.23	-0.06	-0.23	-0.03	-0.14	-0.07	0.00	-0.26	-0.11	-0.27	-0.16
Ro	0.39**	0.36*	-0.03	-0.19	0.05	0.17	0.21	0.21	0.13	0.18	0.19	0.17
Si	0.20	0.14	0.07	0.11	0.17	0.23	0.30	0.32*	0.33*	0.45**	0.28	0.37*
Ya	0.11	0.28	0.21	0.14	0.37**	0.36*	-0.13	0.14	0.11	0.24	0.31*	0.14
Am	0.32	0.12	-0.29	-0.03	0.31	0.12	0.10	0.01	-0.06	0.12	0.03	0.05
Ub	0.38*	0.07	-0.03	-0.03	-0.13	0.10	0.30	0.31	0.16	0.10	0.18	0.33*

	SPEI9						SPEI12					
	Jul	Aug	Sep	Oct	Nov	Mean	Jul	Aug	Sep	Oct	Nov	Mean
Na	0.07	0.14	0.32*	0.21	0.21	0.21	0.15	0.14	0.35*	0.17	0.18	0.23
Bu	-0.15	-0.02	-0.13	-0.06	-0.05	-0.09	-0.29	-0.20	-0.27	-0.13	-0.03	-0.21
Kh	-0.43**	-0.28	-0.45**	-0.36*	-0.24	-0.43**	-0.06	0.13	-0.40*	-0.41**	-0.33*	-0.27
Ma	-0.04	0.05	-0.19	-0.14	-0.21	-0.12	0.10	0.16	-0.28	-0.19	-0.19	-0.09
Su	-0.08	0.03	-0.15	-0.08	-0.19	-0.10	-0.13	-0.05	-0.26	-0.08	-0.16	-0.15
Ro	0.20	0.23	0.08	0.09	0.05	0.13	0.27	0.36*	0.14	0.08	0.07	0.23
Si	0.27	0.31	0.30	0.39**	0.34*	0.34*	0.22	0.33*	0.22	0.38**	0.34*	0.34*
Ya	-0.14	0.15	0.11	0.16	0.14	0.09	-0.19	0.08	0.06	0.13	0.15	0.05

	SPEI9						SPEI12					
	Jul	Aug	Sep	Oct	Nov	Mean	Jul	Aug	Sep	Oct	Nov	Mean
Am	0.00	0.00	-0.08	-0.01	-0.06	-0.03	0.02	-0.12	-0.18	-0.05	-0.04	-0.09
Ub	0.24	0.31*	0.32*	0.29	0.29	0.31	0.28	0.24	0.31	0.26	0.31	0.31

* p-value ≤ 0.10, ** p-value ≤ 0.05

The trends of all climatic variables and SPEI-1 are predominantly increasing. In the Mun River basin, Tmax is rising faster than Tmin, unlike in China (Zhou et al., 2004) and India (Padma Kumari, Londhe, Daniel, & Jadhav, 2007), the first and second largest rice producers, where Tmin has been increasing faster than Tmax. The upward trends in Tmin and Tmax are relatively pronounced in October and November and are more apparent in provinces in the East than in the West. The records also show that in some years, daily Tmax in October in some provinces exceeded 35 °C. If the observed trend continues in the future, the negative effect of Tmax on rice yield is likely to be more visible. The increasing trends in precipitation and SPEI-1 varied considerably. Negative impacts of precipitation on yields mostly occur in September, which is usually the wettest month. In absence of proper drainage, larger volumes of rainwater will likely negatively impact rice development.

The total yield losses due to recent climate trends over the past 30 years (1984-2013) in all provinces are rather low with large uncertainties. When considering only the sensitivity to the temperatures, the yield decreased between 2% to 10% per 1 °C increase in Tmin and Tmax. The rate is similar to the findings by Peng et al. (2004) at the International Rice Research Institute Farm in Philippines who concluded that in the dry season grain yield decreased by approximately 10% for each 1 °C increase in Tmin. The relatively low yield reduction in the Mun River Basin may be explained by the high drought tolerance of the two rice cultivars grown in the area (Bureau of Rice Research and Development (BRRD), n.d.). Additionally, it is partially due to opposing impacts of temperatures and precipitation, and in part because of opposing impacts between Tmin and Tmax. The impacts are determined by the magnitudes of both the effects and trends in temperatures and precipitation. The standardized regression coefficients (Bring, 1994) reveals that changes in precipitation have the most effect on yield change when Tmin, Tmax and precipitation equally changed by one standard deviation. However, in the Mun River Basin yield impacts are driven by trends in temperatures rather than in precipitation, consistent with recent research (Bhatt et al., 2014; Lobell et al., 2011). This is because trends in temperature are stronger than in precipitation.

Irrigation provides water to crops in months with high moisture deficit. Hence, the expectation is that in provinces with a higher rate of irrigated area (Table 2-1) the strength of associations of rice productions with climatic variables such as precipitation and SPEI-1 would be weaker. We observed that in the Nakhon Ratchasima province where almost 20% of paddy fields were irrigated, the relationships were not as strong as in Yasothon province (4.9% irrigated land) where the highest correlations of rice yields with precipitation and with SPEI-1 are found. However, the effect of irrigation on correlation values is not very clearly visible. For example, in Nakhon Ratchasima, Buriram, Maha Sarakham, Surin, Roi Et, and Si Sa Ket, correlations between rice yields and SPEI-1 are comparable while the irrigated land of Nakhon Ratchasima is quite distinct from that of the rests. In the Khon Kaen province, with the least irrigated area at only 0.5%, the strength of association of rice yields with SPEI-1 was weaker than in Yasothon province. We find that in some provinces, positive gains from the precipitation trends compensated for negative impacts from the temperature trends. This means that possible losses in yields due to global warming may be compensated, to some extent, by improving water deficiency through irrigation and crop water management. The role of irrigation in climate – yield relationships deserves further study.

In this study, we disregard the effect of CO_2 levels on rice yields because the year-to-year differences in CO_2 concentration is too small to produce the measurable yield changes (Lobell & Field, 2007). The rice yields increase by 1% for 20 ppm additional of CO_2 concentration (Parry, Rosenzweig, Iglesias, Livermore, & Fischer, 2004). Consequently, since 1984, the CO_2 level increased by 55 ppm (Dlugokencky & Tans, 2017) contributing to roughly 2.8% increase in rice yield. This is three times lower than the total yield losses due to climate trends. Furthermore, Thailand has relatively low CO_2 emission rates (Olivier, Peters, Muntean, & Janssens-Maenhout, 2016), hence we conclude that rice production suffers more from climate change than benefitting from an increase in CO_2 concentration.

3.5 Conclusions

One of the strengths of our study is on the selection of spatial and temporal scales. We carried out our analysis on the provincial level covering all the 10 provinces in the Mun River Basin. We showed that there is large variation in the impacts of climate trends on rice yields across the provinces. For example, among the months with highest correlations, rice yields have gained from the precipitation trends of the past 30 years (1984-2013) in five of the 10 provinces. While, in the other five provinces, rice yields have suffered from the precipitation trends during the same period (Figure 3-4). If the analysis were limited to the basin scale, these variations would be overlooked and the results would be misleading as the gains and losses cancel out. On the temporal scale, we used the monthly time step and showed there are large variations in the degree to which rice yields are affected by the climates of different months of the growing season (Jul. to Nov.). This also allowed us to identify specific months that rice is vulnerable to changes in temperatures and precipitation. Furthermore, we find that the regression models of particular months presented higher explanatory power than that of the seasonal averages. Similarly, we also find that the multiple regression based on Tmin and Tmax separately can explain the yield variance better than Tave alone.

The SPEI-1 presented stronger correlations with rice yields than monthly precipitation and SPEI with higher aggregation periods (e.g. SPEI-3 and SPEI-6). Thus, SPEI-1 may be used as an effective tool for monitoring water stress on rice cultivation in the region. Further studies are necessary to determine the level of SPEI drought severity that should trigger a mitigation action.

Although the impact of past climate change (1984-2013) on rice yields in the Mun River Basin is still relatively low, the yield reduction is likely to be more serious in the future if the observed trends of temperatures and precipitation continue. For example, increasing trends of Tmax in October (reaching Tmax higher than 35°C) may lead to spikelet sterility and severe yield losses. Similarly, the high precipitation intensity in month September, which is one of the major causes of low rice yields in the area, together with increasing trend of precipitation in this month may result in more severe yield losses.

The results of this study provide understanding of the significance of climate impacts on rice yields in the Mun River Basin in Thailand. The opposing impacts of Tmin and Tmax highlight the need for further research regarding effects of global warming on rice yields and its precise physiological mechanism. This study we used only climate variables as influencing factors. More insights may be obtained by including non-climatic factors, such as market prices, agronomic practices, cultivars and fertilizers, in the analysis.

References

Adams, R. M., Hurd, B. H., Lenhart, S., & Leary, N. (1998). Effects of global climate change on agriculture: an interpretative review. *Climate Research, 11*(1), 19-30.

Allen, R. G., Pereira, L. S., Raes, D., & Smith, M. (1998). FAO Irrigation and Drainage Paper No. 56, Crop Evapotranspiration (Guidelines for Computing Crop Water Requirements). FAO. *Water Resources, Development and Management Service, Rome, Italy.*

Babel, M. S., Agarwal, A., Swain, D. K., & Herath, S. (2011). Evaluation of climate change impacts and adaptation measures for rice cultivation in Northeast Thailand. *Climate Research, 46*(2), 137.

Beguería, S., & Vicente-Serrano, S. (2013). SPEI: calculation of the standardised precipitation-evapotranspiration index. *URL:* *https://cran.r-project.org/package=SPEI.*

Beguería, S., Vicente-Serrano, S. M., Reig, F., & Latorre, B. (2014). Standardized precipitation evapotranspiration index (SPEI) revisited: parameter fitting, evapotranspiration models, tools, datasets and drought monitoring. *International Journal of Climatology, 34*(10), 3001-3023.

Bhatt, D., Maskey, S., Babel, M. S., Uhlenbrook, S., & Prasad, K. C. (2014). Climate trends and impacts on crop production in the Koshi River basin of Nepal. *Regional Environmental Change, 14*(4), 1291-1301.

Bring, J. (1994). How to standardize regression coefficients. *The American Statistician, 48*(3), 209-213.

Brisson, N., Gate, P., Gouache, D., Charmet, G., Oury, F.-X., & Huard, F. (2010). Why are wheat yields stagnating in Europe? A comprehensive data analysis for France. *Field crops research, 119*(1), 201-212.

Bureau of Rice Research and Development (BRRD). (n.d., 14 July 2016). Rice Knowledge Bank (in Thai). *Ministry of Agriculture and Cooperatives, Thailand.* Retrieved from http://www.brrd.in.th/rkb/

Challinor, A., Watson, J., Lobell, D., Howden, S., Smith, D., & Chhetri, N. (2014). A meta-analysis of crop yield under climate change and adaptation. *Nature Climate Change, 4*, 287-291.

Dlugokencky, E., & Tans, P. (2017). Trends in Atmospheric Carbon Dioxide. Retrieved from www.esrl.noaa.gov/gmd/ccgg/trends/

Dorman, M., Perevolotsky, A., Sarris, D., & Svoray, T. (2015). The effect of rainfall and competition intensity on forest response to drought: lessons learned from a dry extreme. *Oecologia, 177*(4), 1025-1038.

Edmonds, J. A., & Rosenberg, N. J. (2005). Climate change impacts for the conterminous USA: An integrated assessment summary. In *Climate Change Impacts for the Conterminous USA* (pp. 151-162): Springer.

Erda, L., Wei, X., Hui, J., Yinlong, X., Yue, L., Liping, B., & Liyong, X. (2005). Climate change impacts on crop yield and quality with CO2 fertilization in China. *Philosophical Transactions of the Royal Society B: Biological Sciences, 360*(1463), 2149-2154.

Fischer, G., Shah, M., Tubiello, F. N., & Van Velhuizen, H. (2005). Socio-economic and climate change impacts on agriculture: an integrated assessment, 1990–2080. *Philosophical Transactions of the Royal Society of London B: Biological Sciences, 360*(1463), 2067-2083.

Fox, J. (2015). *Applied regression analysis and generalized linear models*: Sage Publications.

Fuhrer, J., Beniston, M., Fischlin, A., Frei, C., Goyette, S., Jasper, K., & Pfister, C. (2006). Climate risks and their impact on agriculture and forests in Switzerland. In *Climate Variability, Predictability and Climate Risks* (pp. 79-102): Springer.

Guttman, N. B. (1999). Accepting The Standardized Precipitation Index: A Calculation Algorithm1. In: Wiley Online Library.

Hargreaves, G. H. (1994). Defining and using reference evapotranspiration. *Journal of irrigation and drainage engineering, 120*(6), 1132-1139.

Hekstra, G. (1986). Will climatic changes flood the Netherlands? Effects on agriculture, land use and well-being. *Ambio,* 316-326.

Jagadish, S., Sumfleth, K., Howell, G., Redoña, E., Wassmann, R., & Heuer, S. (2010). Temperature effects on rice: significance and possible adaptation. *environments: coping with adverse conditions and creating opportunities,* 19.

Lobell, D. B., Cahill, K. N., & Field, C. B. (2007). Historical effects of temperature and precipitation on California crop yields. *Climatic Change, 81*(2), 187-203.

Lobell, D. B., & Field, C. B. (2007). Global scale climate–crop yield relationships and the impacts of recent warming. *Environmental research letters, 2*(1), 014002.

Lobell, D. B., Ortiz-Monasterio, J. I., Asner, G. P., Matson, P. A., Naylor, R. L., & Falcon, W. P. (2005). Analysis of wheat yield and climatic trends in Mexico. *Field crops research, 94*(2), 250-256.

Lobell, D. B., Schlenker, W., & Costa-Roberts, J. (2011). Climate trends and global crop production since 1980. *Science, 333*(6042), 616-620.

Mavromatis, T. (2007). Drought index evaluation for assessing future wheat production in Greece. *International Journal of Climatology, 27*(7), 911-924.

McKee, T. B., Doesken, N. J., & Kleist, J. (1993). *The relationship of drought frequency and duration to time scales.* Paper presented at the Proceedings of the 8th Conference on Applied Climatology.

Nagarajan, S., Jagadish, S., Prasad, A. H., Thomar, A., Anand, A., Pal, M., & Agarwal, P. (2010). Local climate affects growth, yield and grain quality of aromatic and non-aromatic rice in northwestern India. *Agriculture, Ecosystems & Environment, 138*(3), 274-281.

Nicholls, N. (1997). Increased Australian wheat yield due to recent climate trends. *Nature, 387*(6632), 484-485.

Olivier, J. G. J., Peters, J. A. H. W., Muntean, M., & Janssens-Maenhout, G. (2016). *Trends in global CO2 emissions: 2016 report*: PBL Netherlands Environmental Assessment Agency, European Commission, Joint Research Centre (EC-JRC).

Padma Kumari, B., Londhe, A., Daniel, S., & Jadhav, D. (2007). Observational evidence of solar dimming: Offsetting surface warming over India. *Geophysical Research Letters, 34*(21).

Palmer, W. C. (1965). *Meteorological drought*: US Department of Commerce, Weather Bureau Washington, DC, USA.

Parry, M. L., Rosenzweig, C., Iglesias, A., Livermore, M., & Fischer, G. (2004). Effects of climate change on global food production under SRES emissions and socio-economic scenarios. *Global Environmental Change, 14*(1), 53-67.

Pathak, H., Ladha, J., Aggarwal, P., Peng, S., Das, S., Singh, Y., . . . Sastri, A. (2003). Trends of climatic potential and on-farm yields of rice and wheat in the Indo-Gangetic Plains. *Field crops research, 80*(3), 223-234.

Peng, S., Huang, J., Sheehy, J. E., Laza, R. C., Visperas, R. M., Zhong, X., . . . Cassman, K. G. (2004). Rice yields decline with higher night temperature from global warming. *Proceedings of the National Academy of Sciences of the United States of America, 101*(27), 9971-9975.

Potop, V., Boroneanţ, C., Možný, M., Štěpánek, P., & Skalák, P. (2014). Observed spatiotemporal characteristics of drought on various time scales over the Czech Republic. *Theoretical and applied climatology, 115*(3-4), 563-581.

Rowhani, P., Lobell, D. B., Linderman, M., & Ramankutty, N. (2011). Climate variability and crop production in Tanzania. *Agricultural and Forest Meteorology, 151*(4), 449-460.

Schlenker, W., & Lobell, D. B. (2010). Robust negative impacts of climate change on African agriculture. *Environmental research letters, 5*(1), 014010.

Tan, C., Yang, J., & Li, M. (2015). Temporal-Spatial Variation of Drought Indicated by SPI and SPEI in Ningxia Hui Autonomous Region, China. *Atmosphere, 6*(10), 1399-1421.

Tao, F., Yokozawa, M., Liu, J., & Zhang, Z. (2008). Climate-crop yield relationships at provincial scales in China and the impacts of recent climate trends. *Climate Research, 38*(1), 83-94.

The National Drought Mitigation Center (NDMC). (n.d.). Interpretation of Standardized Precipitation Index Maps. *Moitoring Tools. Climate Division SPI.* Retrieved   from http://drought.unl.edu/MonitoringTools/ClimateDivisionSPI/Interpretation .aspx

Thornthwaite, C. W. (1948). An approach toward a rational classification of climate. *Geographical review*, 55-94.

Tucker, C. J. (1979). Red and photographic infrared linear combinations for monitoring vegetation. *Remote sensing of environment, 8*(2), 127-150.

Vicente-Serrano, S. M., Beguería, S., & López-Moreno, J. I. (2010). A Multiscalar Drought Index Sensitive to Global Warming: The Standardized Precipitation Evapotranspiration Index. *Journal of Climate, 23*(7), 1696-1718. doi:10.1175/2009jcli2909.1

Wang, W., Zhu, Y., Xu, R., & Liu, J. (2015). Drought severity change in China during 1961–2012 indicated by SPI and SPEI. *Natural Hazards, 75*(3), 2437-2451.

Wassmann, R., Jagadish, S., Heuer, S., Ismail, A., Redona, E., Serraj, R., . . . Sumfleth, K. (2009). Climate change affecting rice production: the physiological and agronomic basis for possible adaptation strategies. *Advances in agronomy, 101,* 59-122.

Wei, T., Cherry, T. L., Glomrød, S., & Zhang, T. (2014). Climate change impacts on crop yield: Evidence from China. *Science of the Total Environment, 499,* 133-140.

Welch, J. R., Vincent, J. R., Auffhammer, M., Moya, P. F., Dobermann, A., & Dawe, D. (2010). Rice yields in tropical/subtropical Asia exhibit large but opposing sensitivities to minimum and maximum temperatures. *Proceedings of the National Academy of Sciences, 107*(33), 14562-14567.

Wopereis, M., Defoer, T., Idinoba, P., Diack, S., & Dugué, M. (2008). Participatory learning and action research (PLAR) for integrated rice management (IRM) in inland valleys of sub-Saharan Africa: technical manual. *WARDA Training Series. Africa Rice Center, Cotonou, Benin, 128,* 26-32.

World Meteorological Organization. (1992). *International meteorological vocabulary (2nd ed.)*: WMO.

Yoshida, S., Satake, T., & Mackill, D. (1981). High-temperature stress in rice [study conducted at IRRI, Philippines]. *IRRI Research Paper Series (Philippines).*

Zargar, A., Sadiq, R., Naser, B., & Khan, F. I. (2011). A review of drought indices. *Environmental Reviews, 19*(NA), 333-349. doi:10.1139/a11-013

Zhou, L., Dickinson, R. E., Tian, Y., Fang, J., Li, Q., Kaufmann, R. K., . . . Myneni, R. B. (2004). Evidence for a significant urbanization effect on climate in China. *Proceedings of the National Academy of Sciences of the United States of America, 101*(26), 9540-9544.

4

Flood hazard assessment

This chapter is based on:
Prabnakorn, S., Suryadi, F., Chongwilaikasem, J., & de Fraiture, C. (2019). Development of an integrated flood hazard assessment model for a complex river system: a case study of the Mun River Basin, Thailand. *Modeling Earth Systems and Environment*, doi:10.1007/s40808-019-00634-7.

Abstract

Flooding is the most frequent natural disasters in Thailand, resulting in the loss of life and damage. In this research, we develop an integrated hydrologic and hydraulic model of the Mun River Basin, Thailand, and employ it to predict flood hazard maps at 10, 25, 50, and 100-year return periods. The 'goodness-of-fit' statistics for probability distributions of rainfall data are investigated to select the best-fit distribution, which is used to derive the design rainfall depths for the scenarios (10, 25, 50, and 100-year return periods). The results highlight the diversity of probability distributions even in the basin scale. The model demonstrates satisfactory results, and the flood hazard maps reveal that the extent of flooding is greater on the left bank than the right, and that the flood depths vary mostly between 0 and 4 m. The results also demonstrate that approximately 60% of floodplain inundation is < 1 m, which is mostly observed at the upstream and central areas of the river. The extent of flooding downstream is not as large as the upstream, but generally deeper. Due to the flat topography, the duration of flooding is long, which could damage crop growth and yield, including cities in the flooded areas. This information is essential for the government and authorities to develop flood-control measures and flood management strategies. In this case, planning and regulation of the use of land, and levees to protect the exposed cities are recommended. The integrated model can be further developed for the design of flood mitigation measures and flood forecasting and warning systems. The approach and parameters (both initial and calibrated ranges) can be used to guide future model development and in other basins with similar catchment characteristics, particularly where gauge data are not available.

Keywords: Hydrologic model, SWAT, Hydraulic model, HEC-RAS, flood modeling, flood map

4.1 Introduction

Flooding is the most frequent and devastating natural disaster that affects Thailand. Between 1984 and 2014, Thailand suffered from 66 floods, which caused 48.7 million people affected with total damage of approximately $45 billion (Centre for Research on the Epidemiology of Disaster (CRED), 2015). The losses from flooding tend to increase because of a significant increase in flood frequency and intensity, population growth, and demographic and societal shifts (Changnon, 2008; IPCC, 2012). For example, changes in land use (specifically a high degree of urbanization) will reduce the natural capacity to absorb floodwaters, leading to higher damage. As a result, increasing attention is given to the need for lowering flood casualties and related losses. Identification of flood-prone areas is one of the key solutions in flood mitigation (Sarhadi, Soltani, & Modarres, 2012). Flood hazard maps are an effective tool that allows authorities to prioritize and select flood prevention and mitigation measures (European Parliament, 2007). They highlight potentially inundated areas at different probabilities, complemented by parameters such as flood extents, water depths, etc., and indicate flood intensity (Koivumäki et al., 2010; Van Alphen & Passchier, 2007).

The process of identifying flood-prone areas generally involves flood modeling. Integrating flood modelling is increasingly being demonstrated to be necessary because of the complicated nature of river systems, as well as the interactions of various parameters that define basin characteristics. This includes the sophisticated interactions of different river components and floodplains (Devi, Ganasri, & Dwarakish, 2015; Popescu, Jonoski, Van Andel, Onyari, & Moya Quiroga, 2010). In areas with a complex river system, integrated models, which combine a hydrologic (rainfall-runoff) model with one-dimensional (1D) and two-dimensional (2D) hydraulic models are usually adopted. A hydrologic model is now considered as a crucial and necessary tool for water resources management (Devi et al., 2015), and is usually employed to derive flood peaks by routing flood events between streamflow gauging stations. A hydraulic model, on the other hand, is used to simulate flood propagation in the river based on channel geometry (Blackburn & Hicks, 2002; Popescu et al., 2010).

Flood mapping is of great importance and priority in many countries. For example, in Europe, the European Parliament enacted a new Flood Directive (2007/60/EC) on 23 October 2007. The purpose of the directive is to establish a framework for the assessment and management of flood risks in Europe. In the directive, member states are required to complete flood hazard and risk maps at the river basin district level by 2013, as they are essential tools in the preparation of management plans (de Moel, van Alphen, & Aerts, 2009; European Parliament, 2007). Consequently, almost all European countries have flood hazard maps available (de Moel et al., 2009).

Switzerland, Norway, Turkey, Colombia, Japan, and Taiwan also publicly reported flood hazard maps, as well as the USA, which obtained flood hazard maps through several national flood programs (Bubeck et al., 2017; de Moel et al., 2009; Demir & Kisi, 2016; Doong, Lo, Vojinovic, Lee, & Lee, 2016; Mosquera-Machado & Ahmad, 2007; Van Alphen & Passchier, 2007).

For Thailand, extensive studies have centered on the central plain of Thailand – the Chao Phraya River Basin (including some important provinces in this area) (Cooper, 2014; Kwak, Park, Yorozuya, & Fukami, 2012; Nakmuenwai, Yamazaki, & Liu, 2017; Rakwatin, Sansena, Marjang, & Rungsipanich, 2013; Son, Chen, Chen, & Chang, 2013; Vojinovic et al., 2016). From the studies, flood hazard maps were mostly derived from historic floods and usually from only one event (the 2011 flood is commonly adopted). The maps are thus representative of past flood events in the absence of the probability of occurrences. However, based on the Flood Directive (European Parliament, 2007) conclusion that flood hazard maps should be accompanied by recurrence periods at low, medium, and high probabilities, those maps are not complete. Less attention has been devoted to other parts of the country that are also susceptible to flooding. The Mun River Basin, as an example, is the largest river basin in the northeast, which itself is the most populous region with the most extensive areas of rice cultivation. It is also the only region where the highest quality Thai jasmine rice (a crucial contributor to the Thai economy) can be grown (Kukusamude & Kongsri, 2018; Phoonphongphiphat, 2018). However, neither detailed studies nor flood hazard maps of the basin have been publicly reported.

This paper, therefore, aims to develop an integrated model of the Mun River Basin in Thailand and utilize it to derive flood hazard maps at different scenarios. Integrated modelling is the focus of this study because by using it to simulate the rainfall depths at different probabilities, complete flood hazard maps are obtained. Additionally, it can be used for other purposes in the design and analysis of flood mitigation measures, as well as flood forecasting and warning systems. All procedures involving in model development and calibration are here presented as well as discrepancies between simulated and observed results. The final outcomes (flood hazard maps) are presented and discussed, and possible measures and policies that could potentially mitigate the adverse consequences of flooding are also provided. Furthermore, to achieve the processes mentioned above, the probability distributions of rainfall data and their goodness-of-fit statistics at all gauge stations across the basin are investigated, in order to estimate the rainfall depths at different return periods.

4.2 Data collection

To set up the integrated model, various data sources from multiple agencies are required, as tabulated in Table 4-1. The climatic data are long-term time series from the period of 1985 – 2015, mostly from in-situ gauge stations distributed across the basin, except for solar radiation, which is acquired from the Climate Forecast System Reanalysis (CFSR) dataset - the Global Weather Data for SWAT (Soil and Water Assessment Tool). Missing daily precipitation data were filled through linear regression (Laat, 2012). The equation used is of the form:

$$Y = C + C_1X_1 + C_2X_2 + C_3X_3 + ...,\tag{4.1}$$

where Y is a series of values of the base station, X_i is a series of values of neighboring station i, C is the equation's constant, and C_i is the equation's coefficients. The method is based on fitting a straight line through observations from different sets of neighboring stations. The one with the best coefficient of determination (ρ^2), but not less than 0.5, is accepted for data completion. If a coefficient of determination of 0.5 cannot be achieved, the missing rainfall data is left blank.

Table 4-1 Data description and their sources used in SWAT, rainfall frequency analysis, and HEC-RAS.

Data type	Description	Source
Digital Elevation (DEM)	30 m aggregated to 100 m	Royal Irrigation Department, Thailand
Land use	Vector data exported to 100 m	Land Development Department, Thailand
Soil type and properties	Vector data exported to 100 m and digital files	Land Development Department, Thailand
River network	Vector	Royal Irrigation Department, Thailand
Precipitation	Observed, 53 stations	Royal Irrigation Department and Meteorological Department, Thailand
Temperature	Observed, 18 stations	Meteorological Department, Thailand
Relative humidity	Observed, 14 stations	Meteorological Department, Thailand
Wind speed	Observed, 14 stations	Meteorological Department, Thailand

Data type	Description	Source
Solar radiation	0.25° grid	Global Weather Data for SWAT (https://globalweather.tamu.edu)
River discharge	Observed, 18 stations	Royal Irrigation Department, Thailand
Water level	Observed, 7 stations	Royal Irrigation Department, Thailand
Cross section	Field survey data	Royal Irrigation Department, Thailand

4.3 Methodology

As mentioned previously, the study had, as its aim, the development of an integrated model for the Mun River Basin, which can be subsequently used for flood hazard mapping and further developed for flood forecasting and planning. Thus, all activities relevant to the model development and its important features are described as in Figure 4-1, which can facilitate studies of other basins in Thailand and elsewhere. The model sequentially combines the hydrologic model (SWAT) with a 1D2D hydraulic model (HEC-RAS). The process starts with the development of the hydrologic model using SWAT. The SWAT model was set up for the whole basin and calibrated at all gauge stations presented in Figure 4-2 simultaneously, because of data limitation. The observed discharge data are available only for some tributaries; thus, it is impossible for detailed models at other smaller scales. The validated model was used to simulate runoff from different scenarios, and the computed discharge hydrographs were subsequently fed into the calibrated HEC-RAS model at all connection points (Figure 4-2) as lateral inflows, by the use of HEC Data Storage System (HEC-DSS). At this step, model codification to support the connectivity between the two models was needed to relate features across the models spatially. The whole suite of models employed to support full integration is comprised of SWAT, SWAT-CUP, HEC-GeoRAS, HEC-RAS, and HEC-DSS.

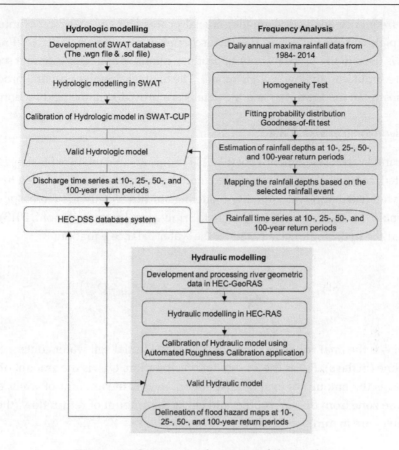

Figure 4-1 Summarized process of the study

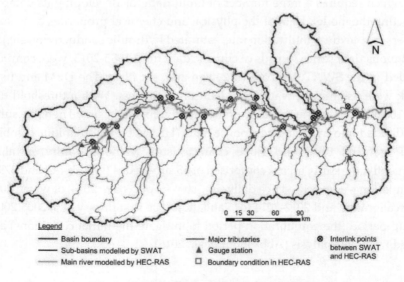

Figure 4-2 Interlinkage between models

Rainfall frequency analysis is another crucial process for the mapping of inundation of floodplains in places where gauge discharge is poor or unavailable. Frequency analysis is used in the present study to produce rainfall depths for different scenarios, which are input into the hydrologic model. The analysis involves several procedures, as demonstrated in Figure 4-1, and the details are provided in the next section.

4.3.1 Hydrologic model: SWAT

The Soil and Water Assessment Tool (SWAT) program, an ArcGIS extension, is a semi-distributed, continuous, processed-based model, which has been employed to support various watershed and water-quality modeling studies worldwide (Abbaspour et al., 2015; Arnold, Moriasi, et al., 2012; Winchell, Srinivasan, Luzio, & Arnold, 2013). SWAT simulated the hydrologic cycle based on the water balance equation:

$$SW_t = SW_0 + \sum_{i=1}^{t} \left(R_{day} - Q_{surf} - E_a - w_{seep} - Q_{gw} \right),   \tag{4.2}$$

where SW_t is the final soil water content, SW_0 is the initial soil water content on day i, t is the time (in days), R_{day} is the amount of precipitation, Q_{surf} is the amount of surface runoff, E_a is the amount of evapotranspiration, w_{seep} is the amount of water entering the vadose zone from the soil profile, and Q_{gw} is the amount of return flow (the unit of all variables are in mm).

In this study, the ArcSWAT 2012 interface is adopted to set up and parameterize the model, which requires a large number of input data for developing its database and constructing the model. Data of the physical and chemical properties of all soil types and layers (i.e., texture, infiltration rate, saturated hydraulic conductivity, etc.), as well as continuous time-series records of climatic data from 1985-2015, were compiled and embedded in the SWAT database (the .wgn and .sol files). The DEM and its stream network were used in the watershed delineation process. With a threshold drainage area of 100,000 ha, the watershed is first divided into sub-basins. Then, the sub-basins are subdivided into hydrologic response units (HRUs) based on unique combinations of land use, soil type, and slope characteristics (Arnold, Kiniry, et al., 2012). Consequently, the study area is comprised of 63 sub-basins (Figure 4-2) and 379 HRUs. The simulation was executed at a daily time step from 2005 to 2014, of which 2008-2011 was for calibration and 2012-2014 for validation. The first three years, 2005-2008, are a warm-up period (the equilibration period to mitigate the initial conditions) and was excluded from the analysis (Abbaspour et al., 2015).

The model's ability to simulate the hydrologic process and applicability in the specified watershed was evaluated during the calibration and validation process using an automated calibration called SWAT-CUP (SWAT Calibration Uncertainty Programs) (Abbaspour, 2015) with the SUFI-2 (Sequential Uncertainty Fitting) algorithm (Abbaspour, Johnson, & Van Genuchten, 2004). This algorithm is widely used to estimate parameter ranges with all relevant uncertainties (input variables, conceptual model, observed data, parameters, etc.) (Abbaspour et al., 2015). In our case, all 18 discharge outlets over the entire basin were parameterized and optimized simultaneously. Fourteen parameters (Table 4-2) and their initial ranges were selected and assigned based on the "Training Manual: ArcSWAT 2012" (Mekong River Commission Secretariat, 2014), which recommends some initial parameters' ranges for basins in Thailand nearby the Mekong River supplemented with review literature. Once the model was parameterized, and the ranges were assigned, an iteration on a daily basis began.

Table 4-2 Fourteen SWAT model parameters, methods, initial values, and their calibrated ranges and values.

No	Parameter	Method	Initial values	Initial ranges	Calibrated ranges	Best calibrated values
1	ALPHA_BF	Replace	0.048	0 to 1	0.16 to 0.49	0.20
2	CANMX	Replace	0	0 to 10	2.75 to 8.25	4.33
3	CH_K2	Replace	0	-0.01 to 10	-0.01 to 3.4	0.16
4	CH_N2	Replace	0.014	0.014 to 0.3	0.14 to 0.3	0.29
5	CN2	Multiply	66 to 87	-0.47 to 0.13	-0.47 to -0.25	-0.44
6	EPCO	Replace	1	0 to 1	0.44 to 0.81	0.74
7	ESCO	Replace	0.95	0 to 1	0.66 to 0.85	0.67
8	GWQMN	Replace	1000	0 to 5000	1033.92 to 3102.92	1325.65
9	GW_DELAY	Replace	31	0 to 500	0 to 140.08	13.31
10	GW_REVAP	Replace	0.02	0.02 to 0.2	0.12 to 0.17	0.16
11	REVAPMN	Replace	750	0 to 500	118.05 to 354.26	147.10
12	SOL_AWC	Added	0-0.35	0 to 0.65	0.29 to 0.58	0.43
13	SURLAG	Replace	4	0.05 to 24	3.65 to 7.27	6.99
14	OV_N	Replace	0.05 to 0.14	0.01 to 30	5.53 to 16.58	8.39

The model performance and efficiency were subsequently assessed using three indicators: coefficient of determination (R^2), Coefficient of efficiency (or Nash-Sutcliffe (NS) Nash and Sutcliffe (1970)), and the percent bias (PBIAS). The R^2 and NS are the most widely used statistic documented for calibration and validation (Arnold, Moriasi, et al., 2012). R^2 defines how well the simulated values explain the observed dispersion. Its range varies from 0 to 1, with 0 indicating no correlation at all and 1 representing the perfect correlation between simulated and observed values (Krause, Boyle, & Bäse, 2005). Typically, for values of $R^2 \geq 0.5$, the simulated results are considered "acceptable" (Moriasi et al., 2007). The NS measures the proportion of observed flow variance that is replicated by the model. Its range lies between $-\infty$ and 1, from the worst to perfect fit. Simulated results are considered "good" for NS $\geq$ 0.75, "satisfactory" for $0.36 \leq$ NS $<$ 0.75, and "unsatisfactory" for NS $<$ 0.36 (Motovilov, Gottschalk, Engeland, & Rodhe, 1999; W. Van Liew, G. Arnold, & D. Garbrecht, 2003). If NS $\leq$ 0, the mean of observed values is a more accurate predictor than simulated values (Arnold, Moriasi, et al., 2012). The PBIAS expresses the average propensity of the simulated values to be larger, equal, or smaller than their observed ones. The ideal value of PBIAS is zero; positive values reveal a model bias towards underestimation, while negative values reveal the opposite (Gupta, Sorooshian, & Yapo, 1999). The classification standard by Van Liew, Veith, Bosch, and Arnold (2007) was adopted in this study; the model performance is considered "very good" for $|$PBIAS$|$ $<$ 10%, "good" for 10% $\leq$ $|$PBIAS$|$ $<$ 15%, "satisfactory" for 15% $\leq$ $|$PBIAS$|$ $<$ 25%, and "unsatisfactory" for $|$PBIAS$|$ $\geq$ 25%. When the model reaches a satisfactory performance, the model with the calibrated parameters is used to compute runoff for four rainfall scenarios: 10, 25, 50, and 100-year return periods.

4.3.2 Hydraulic model – HEC-RAS

The Hydrologic Engineering Center's River Analysis System (HEC-RAS 5.0.6) software was employed to simulate flood propagation in the Mun River and over its floodplain. The combined one-dimensional and two-dimensional hydraulic (1D2D) model is governed by the conservation of continuity (4.3) and the conservation of momentum (4.4).

$$\frac{\partial A_T}{\partial t} + \frac{\partial Q}{\partial x} - q_1 = 0, \tag{4.3}$$

$$\frac{\partial Q}{\partial t} + \frac{\partial QV}{\partial x} + gA\left(\frac{\partial z}{\partial x} + \frac{Q|Q|n^2}{2.208R^{4/3}A^2}\right) = 0 \tag{4.4}$$

Where Q is the flow, A_T is the total flow area, t is the time, x is the distance along the channel, q_1 is the lateral inflow per unit length, V is the average velocity, g is the gravitational acceleration, A is the cross-sectional area, z is the elevation of the water surface, R is the hydraulic radius and n is the Manning friction coefficient. The continuity and momentum equations for the channel and floodplain are simplified under the assumption that a water surface at each cross-section is horizontally stable across both adjacent floodplains. Thus, the exchange of momentum between the main channel and its floodplains can be neglected. These unsteady flow equations are estimated using implicit finite differences and solved numerically by the four-point implicit scheme (box scheme) (Brunner, 2016a).

A 1D river network model of the Mun River was first established by using HEC-GeoRAS and integrated with the floodplains to allow the interaction of the flows in the river and over the floodplains. With a total length of 733 km, the river scheme consists of approximately 900 cross-sections at about 800-m intervals. They were interpolated from 9 field survey cross-sections available in the area. The geometric data of stream centerline, flow path, channel bed and overbank boundaries (of the cross-section), were digitized from a digital terrain model (TIN format) before being imported to HEC-RAS. Then, the first Manning's n values were set (they were calibrated later), and a daily flow hydrograph and a daily stage (water level) hydrograph were input as upstream and downstream boundary conditions, respectively, in the form of HEC-DSS files. All lateral inflow hydrographs from the hydrologic model representing discharges of all tributaries and the discharge of the Chi River were incorporated into the main channel as point sources throughout the river. Five gauges were included for model calibration and verification. The 1D model was connected to the floodplain by a weir type structure at the river banks, and a 2D computational mesh was created at 500 x 500 m grid size. Although the cell size is rather large, considerable hydraulic details are still retained within a cell using the 2D Geometric Preprocessor. The algorithm preprocesses cells, and cell faces to develop detailed hydraulic property tables (elevation versus wetted perimeter, elevation versus area, roughness, etc.) based on the underlying terrain (5 x 5 m in this case). As such, HEC-RAS can produce detailed results (for example, a cell can be partially wet), which is an advantage over other models that use a single elevation for each cell (Brunner, 2016b). The computational interval is 3 minutes, and the intervals of mapping, hydrographs, and detailed outputs are 1 hour.

The hydraulic model of the Mun River was calibrated using the 'Automated Roughness Calibration' procedure in HEC-RAS. The method performs an automated calibration of the Manning's n values, often the most uncertain and variable input in an unsteady flow model. The river was first broken up into smaller segments between five gauges. Then the table of various flow rates (low to high) versus roughness factors for each segment was established before the automated calibration of all those segments starts. By setting the Optimization Method to Global, the procedure begins calibrating the roughness values at all segments simultaneously. This is done by adjusting the roughness factors along the segment to follow the observed discharges or observed stages at the gauge upstream of that segment. The observed stages were selected for calibrations at all five gauges in our case because it is the most accurate and reliable hydrologic data (Brunner, 2016c). The computation was conducted over five months (July to November) of the year 2014 for calibration and of the year 2012 for verification. The model performance was evaluated before scenario analyses using three indicators: R^2, NS, and the Mean absolute error (MAE). The calibrated model with acceptable performance is further used to simulated design flood scenarios at 10, 25, 50, and 100-year return periods.

4.3.3 Rainfall frequency analysis

Rainfall frequency analysis is a statistical approach to anticipate rainfall depths of extreme events based on the optimal probability for a specific location corresponding to different durations. It was carried out to develop four scenarios: 10, 25, 50, and 100-year return periods. The method assumes that time series data are stationary without trends. Thus the future time series is deemed to reveal similar frequency distribution to the historical observations (Raes, Willems, & Gbaguidi, 2006). The analysis involves the steps shown in Figure 4-1. Homogeneity tests are used to ensure that rainfall records are from the same population, which is homogenous and independent. Here, we performed a homogeneity test based on the cumulative deviations from the mean (Buishand, 1982). Considering at a significant level of 95% or higher, stations with homogeneous rainfall records were subsequently fitted to eight probability distributions commonly used in extreme rainfall analysis: Normal or Gaussian (N), Log-Normal (LN), Log-Pearson Type 3 (LP3), Exponential (E), Gumbel (extreme value type 1) (GUM), Generalized Extreme Value (GEV), Weibull (extreme value type 3) (W), and Generalized Pareto (GP) (Alam, Emura, Farnham, & Yuan, 2018). The best fit distribution was determined from the lowest sum of the rank scores of the three goodness-of-fit statistics: Kolmogorov-Smirnov (K-S), Anderson-Daring (A-D), and Chi-Squared (χ^2) tests. Parameters of the best fit distribution were then estimated using the method of moments or the L-moments or the Maximum Likelihood Estimation depending on convenience and suitability to the distributions.

The best fit distribution was then used to estimate daily rainfall depths at 10, 25, 50, and 100-year return periods at each station. They are a single value of rainfall data, while a continuous time series are required for rainfall-runoff simulation. We thus constructed a rainfall hyetograph with a time scale of 5 months (July to November) based on the observed hyetographs of previous floods using the algorithm developed by Tingsanchali and Karim (2010). The design hyetographs of all the rainfall stations for the specified return periods were input in the calibrated hydrologic model.

4.4 Results and discussions

4.4.1 Rainfall analysis

The annual maxima of daily rainfall data at all stations passed the homogeneity test at the significance level 90% or higher, and their mean, standard deviation (SD), skewness and best-fit statistic results are illustrated in Table 4-3. Among all tested probability distributions, the GEV is the most suitable distribution based on the K-S, A-D, and χ^2. The result differs from that in the Yom River Basin in northern Thailand, and the Chi River Basin - located just above the Mun River Basin, of which the GUM is the best fit distribution (Kuntiyawichai, 2012; Tingsanchali & Karim, 2010). Additionally, the western provinces produce a wide diversity of probability distributions than the east. This highlights the variety of probability distributions of the rainfall, even at the basin scale. Thus, we agree with Raes et al. (2006) that it is vital to examine the goodness-of-fit statistics of the assumed distributions before design rainfall analysis, and the use of a presumed distribution from another basin, even from the adjacent basin, should be avoided if there is no proof of climatic and catchment similarity. Although the GEV is not the best fit distribution at some stations (10 out of the total 34 stations), it still fits the rainfall at the significance level 95% or higher. Thus, it is selected to estimate the design rainfall depths at 10, 25, 50, and 100-year return periods, and the findings are tabulated in Table 4-4.

Table 4-3 Statistical results and best fit distributions of annual maxima of daily rainfall at all rain-gage stations at the Mun River Basin.

No	Station	Mean	SD	Skewness	Best-Fit Test Statistic Results			Best-Fit Distribution (Sum of Ranks)
					K-S	A-D	χ^2	
1	Na25062	73.95	23.44	0.31	0.079 (GP)	0.238 (GEV)	0.420 (LP3)	GEV, LP3 (6)
2	Na25212	90.33	31.39	1.46	0.099 (GEV)	0.282 (GEV)	3.041 (GUM)	GEV (4)
3	Na25013	85.39	21.90	0.45	0.099 (GEV)	0.413 (GEV)	0.947 (GEV)	GEV (3)
4	Na25093	84.36	24.54	0.53	0.088 (GP)	0.312 (GEV)	1.136 (GEV)	GEV (4)
5	Na25112	67.42	22.62	0.40	0.103 (GUM)	0.281 (GEV)	0.944 (LP3)	GUM (5)
6	Na25042	80.98	30.91	1.69	0.098 (LP3)	0.495 (GEV)	0.055 (E)	GEV (5)
7	Ko14033	82.76	21.77	1.21	0.073 (GEV)	0.140 (LP3)	0.272 (LP3)	LP3 (5)
8	Na25222	87.07	26.16	0.78	0.120 (GP)	0.559 (GEV)	3.756 (N)	GEV (7)
9	Na25602	76.45	15.98	0.43	0.113 (N)	0.327 (LP3)	0.229 (GUM)	N (8)
10	Na25182	81.97	24.03	1.04	0.064 (LP3)	0.149 (GEV)	0.074 (LP3)	LP3 (4)
11	Bu02033	94.06	20.74	0.13	0.063 (LN)	0.177 (GEV)	0.401 (LP3)	LP3 (5)
12	Bu02092	84.18	26.75	1.70	0.074 (W)	0.254 (GEV)	0.960 (W)	GEV (7)
13	Ma21252	90.41	30.27	1.28	0.067 (LP3)	0.143 (GEV)	0.291 (LP3)	LP3 (4)
14	Bu02012	84.38	27.47	1.02	0.092 (GEV)	0.247 (GEV)	0.747 (GEV)	GEV (3)
15	Ma21063	95.05	23.94	0.17	0.114 (GP)	0.525 (GEV)	0.373 (GEV)	GEV (4)
16	Bu02052	97.90	33.35	1.21	0.102 (GEV)	0.278 (GEV)	0.955 (LN)	GEV, LN (6)
17	Su62092	96.33	21.30	0.34	0.084 (GEV)	0.230 (GEV)	0.159 (LN)	LN (7)
18	Su62013	103.12	41.17	2.74	0.100 (GP)	0.317 (GEV)	0.313 (W)	GEV (7)
19	Ro49312	91.36	32.41	0.65	0.063 (LN)	0.139 (GEV)	0.367 (GUM)	LN (6)
20	Su62043	100.46	29.31	0.78	0.081 (GP)	0.241 (GEV)	0.347 (LN)	GEV (6)
21	Su62052	97.41	32.40	0.69	0.065 (GEV)	0.163 (GEV)	0.418 (W)	GEV (6)
22	Su62112	97.34	29.64	0.62	0.076 (LN)	0.207 (GEV)	0.464 (GEV)	GEV, LP3 (6)
23	Si57052	110.21	37.94	0.37	0.087 (GEV)	0.212 (GEV)	0.019 (GUM)	GEV (8)
24	Si57072	110.20	37.41	1.19	0.085 (GP)	0.339 (GEV)	0.474 (GEV)	GEV (4)
25	Si57082	92.67	30.19	1.17	0.084 (GEV)	0.278 (GEV)	0.343 (GEV)	GEV (3)
26	Si57203	111.55	40.74	1.95	0.070 (GEV)	0.182 (GEV)	0.605 (LP3)	GEV (4)
27	Am76012	102.14	30.33	0.25	0.068 (GP)	0.321 (GEV)	0.962 (N)	GEV (6)
28	Si57063	96.47	30.84	0.97	0.081 (GEV)	0.208 (GEV)	0.044 (W)	GEV (4)

No	Station	Mean	SD	Skewness	Best-Fit Test Statistic Results			Best-Fit Distribution (Sum of Ranks)
					K-S	A-D	χ^2	
29	Am76032	103.81	29.80	2.09	0.088 (W)	0.451 (GEV)	0.464 (LN)	GEV (5)
30	Ub67013	108.59	32.62	0.29	0.087 (LN)	0.153 (GEV)	0.542 (LN)	LN (5)
31	Ub67092	105.08	38.77	2.03	0.066 (GEV)	0.187 (GEV)	0.216 (W)	GEV (6)
32	Ub67132	101.04	25.23	0.66	0.088 (GUM)	0.211 (LN)	1.036 (GUM)	GUM (6)
33	Ub67112	115.68	36.08	0.32	0.124 (GP)	0.956 (LP3)	0.440 (W)	GEV, W (9)
34	Ub67022	122.01	48.59	1.83	0.074 (LP3)	0.183 (GEV)	0.461 (GEV)	GEV (4)

The stations are listed by from west to east.

Table 4-4   Three parameters of the GEV distribution and the rainfall depths at 10, 25, 50, and 100-year return periods at all stations.

No	Station	Parameter			Daily rainfall depth (mm) at return period			
		shape	scale	location				
		k	σ	μ	10 years	25 years	50 years	100 years
1	Na25062	-0.14	21.93	63.97	106.36	120.61	130.04	138.53
2	Na25212	0.18	19.58	74.74	129.28	159.91	186.34	216.16
3	Na25013	-0.05	19.04	75.35	115.75	131.39	142.49	153.11
4	Na25093	-0.09	21.79	73.54	118.01	134.26	145.46	155.91
5	Na25112	-0.10	20.08	57.67	98.08	112.53	122.39	131.51
6	Na25042	0.23	17.29	65.98	116.93	147.65	175.18	207.29
7	Ko14033	0.07	15.97	72.34	111.30	129.68	144.14	159.22
8	Na25222	0.08	19.38	74.19	122.07	145.04	163.26	182.42
9	Na25602	-0.16	14.62	70.05	97.62	106.55	112.34	117.47
10	Na25182	0.07	17.81	70.37	113.79	134.25	150.34	167.12
11	Bu02033	-0.22	20.48	85.96	122.32	133.02	139.63	145.26
12	Bu02092	0.07	18.76	72.07	117.54	138.79	155.43	172.71
13	Ma21252	0.13	20.55	75.55	129.24	157.00	179.91	204.81
14	Bu02012	0.09	19.89	71.01	120.52	144.51	163.64	183.84
15	Ma21063	-0.20	23.35	85.51	127.77	140.59	148.63	155.56
16	Bu02052	-0.03	26.76	83.21	141.48	164.90	181.85	198.34
17	Su62092	-0.15	19.91	87.37	125.57	138.28	146.64	154.13
18	Su62013	0.20	22.69	84.61	148.82	185.60	217.67	254.20
19	Ro49312	-0.03	27.31	76.49	135.65	159.25	176.26	192.75
20	Su62043	0.03	23.20	86.32	140.43	164.40	182.65	201.17
21	Su62052	-0.02	27.09	82.41	141.75	165.80	183.29	200.37
22	Su62112	-0.03	25.01	83.66	137.98	159.73	175.45	190.71
23	Si57052	-0.14	35.25	94.07	162.47	185.60	200.95	214.80
24	Si57072	0.22	22.50	90.96	156.63	195.75	230.58	270.98
25	Si57082	0.11	20.96	78.01	131.60	158.51	180.39	203.86
26	Si57203	0.15	25.66	92.44	160.81	197.07	227.41	260.78
27	Am76012	-0.17	29.04	89.54	144.04	161.56	172.88	182.87
28	Si57063	-0.03	25.41	82.50	137.89	160.19	176.34	192.07
29	Am76032	-0.02	21.84	91.70	139.58	159.01	173.15	186.96
30	Ub67013	-0.16	30.93	95.02	153.48	172.47	184.81	195.76
31	Ub67092	0.15	24.29	86.84	151.91	186.67	215.87	248.08
32	Ub67132	-0.08	21.77	90.07	134.93	151.56	163.12	173.97
33	Ub67112	-0.10	32.39	99.96	165.13	188.44	204.34	219.05
34	Ub67022	0.21	28.12	98.45	179.44	226.89	268.76	316.93

The stations are listed by from west to east.

4.4.2 Hydrologic model calibration and validation

The SWAT simulation can capture the overall dynamics of the flow in the Mun River Basin, except for a few stations (Figure 4-3). This is because all 18 discharge outlets were parameterized and optimized simultaneously, so some outlets (e.g., M.159 and M.176) may individually perform poorly when an overall satisfactory simulation is achieved (Abbaspour et al., 2015). The stations M.2A, M.184, and M.186 are slightly downstream of four important dams: Lam Takhong, Lam Phra Phloeng, Mun Bon, and Lam Chae, and the flow regimes at these outlets are not well simulated (e.g., Figure 4-4a) because the operation of the dams completely controls their flow regimes.

We also developed a SWAT model incorporating the dam operation; however, the results are not significantly different from the one without the dams, particularly at the outlets after station M.184 further downstream. There are three possible reasons for this. The first is that the outflows (in m³/s) from the reservoirs used in the SWAT model were derived from the total daily flow (in million m³) reported by the Royal Irrigation Department of Thailand (2018) due to lack of accurate flow-rate data. The calculation assumed equal release of water during the day. As a result, some mismatches in amounts and timing are observed. The second reason is that the areas along the course of the river, and its tributaries downstream of the dams, are highly utilized for agriculture, due to which slight differences in the flow may not be detected. The third reason could be that after station M.184, no influence from the dams is observed on the flow of the main river. The model surprisingly well simulated the flow dynamics at M.164 (R^2 = 0.76, NS = 0.74 and PBIAS = 15.9% for calibration and R^2 = 0.52, NS = 0.43 and PBIAS = 46.2% for validation) and M.185 (R^2 = 0.76, NS = 0.61 and PBIAS = 36.1% for calibration and R^2 = 0.89, NS = 0.89 and PBIAS = -4.9% for validation). This is in spite of the fact that the stations are situated slightly downstream of the Lam Takhong Dam, and Lam Nang Rong Dam, respectively.

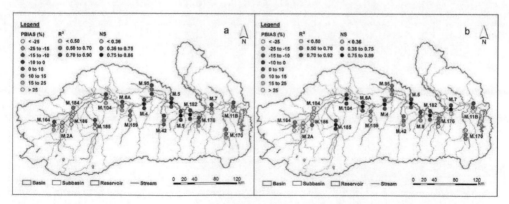

Figure 4-3 R^2, NS, PBIAS, and results from model calibration (a) and validation (b)

The R^2, NS, and PBIAS at the outlets along the main Mun River (M.104, M.6A, M.4, M.5, M.182) are excellent (Figure 4-3, Figure 4-4b, and Figure 4-4c). However, outlets M.7 (Figure 4-4d) and M.11B have large positive PBIAS values indicating underestimated discharges. This is because these last two outlets are located downstream of the confluence of the Mun and Chi Rivers (Figure 2-1c), where the observed discharges accounted for the total flow from both rivers. However, the model results here represent only the discharges from the Mun River. Further, the small forward shifts in simulated hydrographs (particularly at peak flows Figure 4-4) are attributed to: small shifts in rainfall data that govern the outlets, the river geometry, and model simplification. The model was developed based on the DEM with a spatial resolution of 100 m. This cannot capture overall dynamics of the flow of the Mun River and its tributaries, which meander with a complex river system (Figure 2-1b). Such simplicity may result in faster movement of the flow than the observed discharges.

Overall, the SWAT model produces satisfactory results for a large basin with complex river geometry and branching. The advantage of calibrating the hydrologic model of the entire basin simultaneously is that the model also provides the runoff data for ungauged tributaries (subbasins) which have noticeable impacts on the main channel routing in the hydraulic model. This cannot be achieved if each sub-basin was modeled and calibrated individually.

Availability and quality of input data are a limitation in this study that influences model performance. We noticed that the outlets with a lot of missing discharge data usually result in poor results. For example, station M.159 has about 33% of missing discharges for validation and the model ends up with bad results (R^2 = 0.52, NS = - 0.42, and PBIAS = -56.7%), while all calibrated results present quite good simulation (R^2 = 0.65, NS = 0.59, and PBIAS = -25.7%) when it has completed data.

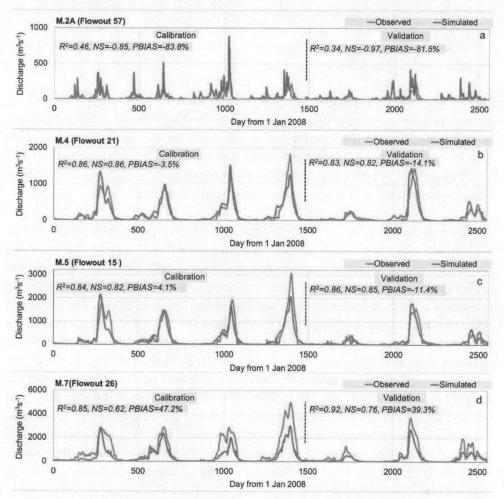

Figure 4-4 Model results of 4 selective gauge stations (out of the total 18 stations). The figure covers both calibration and validation periods presented along with R², NS, and PBIAS

4.4.3 Hydraulic model calibration and validation

Figure 4-5 demonstrates the calibrated and validated results at selected gauge stations (M.104 upstream, and M.7 downstream) by a comparison between measured (observed) and simulated water levels (stage, meters above mean sea level (msl)). In general, the model produces acceptable outcomes, particularly at the gauges downstream (M.7), which present a better flood pattern, flood peak, and flood timing than the ones upstream. This may be because: 1) the ungauged sub-basins are mostly located upstream, which need to use simulated hydrographs instead of observed hydrographs for calibration. 2) There are wide variations in river bathymetry at upstream stations, yet less field survey cross-section data is available compared to the downstream ones. Such variations with fewer precise cross-section data can be a crucial source of error and model instability. 3) Fewer influences from dam and weir operations on gauges downstream.

The R^2 and NS at all gauges reveal promising results representing a reasonable agreement between simulated and observed stage hydrographs. The forward shifts in flood timing (e.g., M.104) correspond to the shifts in the flow hydrographs, as mentioned in Section 4.4.2, together with the reasons above. The MAE varies from 0.20 – 0.70 m and the differences in peak stages are between 0.05 - 0.75 m, most of them being overestimated, which is acceptable for emphasizing flood events.

Table 4-5 demonstrates the calibrated Manning's n values along the Mun River. The roughness coefficient generally deceases with increased flow. However, this trend can reverse if the floodplain is rougher than the channel bottom. This is because the calibrated roughness coefficients are the composite n value encompassing the roughness of both the main channel and its banks (Brunner, 2016c). The calibrated roughness coefficients are close to the roughness of the Chi River, another main river in northeast Thailand (Kuntiyawichai, 2008, 2012), and that of the Yom River in northern Thailand (Tingsanchali & Karim, 2010). Furthermore, the ranges of the roughness coefficients also agree with the values proposed by Chow (1959), which are adopted by many researchers. The calibrated roughness values of the Mun River can be used as a guide in selecting the roughness value for similar stream conditions, particularly where gauge data are not available. This might be valid when transitioning to other hydraulic models, i.e., FLDWAV, MIKE, etc. as was the case for the NOAA National Weather Service (NWS) (Moreda, Gutierrez, Reed, & Aschwanden, 2009).

Table 4-5   Manning's n value along the Mun River

River segment	Manning's n value
Upstream boundary condition to M.104	0.0408 - 0.0426
M.104 – M.6A	0.036
M.6A – M.4	0.0426
M.4 – M.5	0.0258 - 0.0411
M.5 – M.7	0.0249 - 0.045
M.7 – Downstream boundary condition	0.0159 - 0.0315

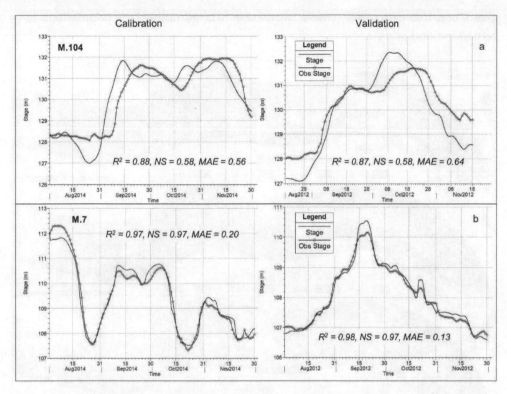

Figure 4-5 Calibrated (left) and validated (right) results of two selective gauge stations (out of the total five stations) presented along with R², NS, and MAE

4.4.4 Flood maps

Figure 4-6 demonstrates the flood hazard maps at 10 (a), 25 (b), 50 (c), and 100-year (d) return periods. The average flood depths are 1.67, 1.74, 1.90, and 1.98 m and the total inundated areas are 215,333; 255,576; 288,084; and 324,239 ha, for 10, 25, 50, and 100-year return periods, respectively. The flood extent expands more on the left bank than the right as a result of flatter terrain. The change in flooded areas declines with increasing recurrence intervals. Approximately 91%, 90%, 88%, and 87% of the total inundated areas have flood depths between 0 and 4 m, and about 60% of floodplain inundation is less than or equal to 1 m, which mostly occurs at the upstream portion of the Mun River. This may be because the river upstream is meandering and complex, which creates natural obstructions to the water flow, inducing higher water levels, and flood impacts. However, it should be noted that the inundated area upstream may be overestimated due to the overestimation of the runoff, as mentioned in Section 4.4.2. At the river downstream, the flood extent is not as large as upstream, but the flood depth is deeper, which may cause adverse impacts in the main districts of Si Sa Ket and Ubon Ratchathani provinces, where are a large number of populations have settled in flood-prone areas.

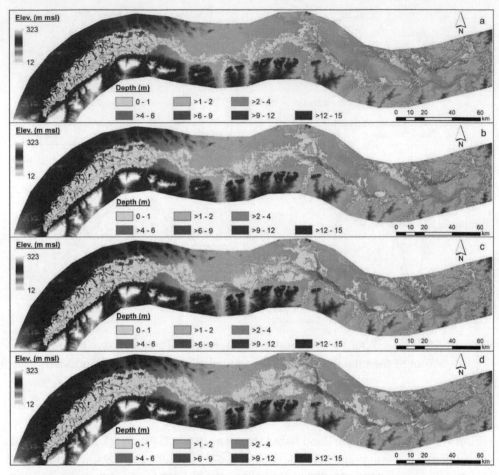

Figure 4-6 Flood map for 10 (a), 25 (b), 50 (c), and 100-year (d) return periods

Table 4-6 illustrates the areas in each land-use class that are affected by different probabilistic scenarios. The flooded areas and percentages are derived by overlaying the flood hazard maps with land-use datasets. Rice fields are most affected because rice is a staple crop of the country occupying the majority of the area in this region. As rice requires much more water than other plants, its plots generally lie on river banks for convenient and economical irrigation, which relies mainly on surface water. Consequently, the fields, particularly at the upstream and the central parts of the basin, are susceptible to flooding. As this area is rather flat, with a bed slope roughly between 0.00007 and 0.00014, the velocity of the water overflows on floodplains is slow, resulting in a long duration of flooding which can cause severe damage to the crop growth and yields. Many cities, for example, Phibun Mangsahan, Mueang, Warin Chamrap districts in Ubon Ratchathani province, Rasi Salai district in Si Sa Ket province, Rattanaburi, Tha Tum, and Chumphon Buri districts in Surin province, and Sateuk district in Buri Ram province are affected. These cities are situated in flood-prone areas which have been preferred for settlement since ancient times, owing to favorable geographic environments which facilitate economic growth, such as easy accessibility and nutrient-rich soils (Douben, 2006). The areas have a high degree of urbanization with high values of assets and infrastructure, thus flooding in this area can cause significant damage.

Table 4-6 Areas of land use (ha) and their percentages (in parentheses) affected by floods at different probabilistic scenarios.

Land use	Return period			
	10-Year	25-Year	50-Year	100-Year
Forest	5,783 (2.7%)	7,383 (2.9%)	8,529 (3.0%)	9,410 (2.9%)
Rice	121,524 (56.4%)	152,143 (59.5%)	179,098 (62.2%)	209,637 (64.7%)
Field crops	4,131 (1.9%)	4,820 (1.9%)	5,423 (1.9%)	6,210 (1.9%)
Perennial crops	12,518 (5.8%)	13,811 (5.4%)	14,520 (5.0%)	15,096 (4.7%)
Grass, pasture, and shrubland	10,910 (5.1%)	12,518 (4.9%)	13,351 (4.6%)	14,144 (4.4%)
Fish farm, Marsh and Swamp	29,588 (13.7%)	30,804 (12.1%)	31,244 (10.9%)	31,568 (9.7%)
Urban area	2,980 (1.4%)	4,338 (1.7%)	5,190 (1.8%)	6,505 (2.0%)
Water bodies (lake, farm pond, etc.)	26,057 (12.1%)	27,840 (10.9%)	28,732 (10.0%)	29,610 (9.1%)
Others (sand pit, abandoned mine, etc.)	1,842 (0.9%)	1,919 (0.8%)	1,997 (0.7%)	2,059 (0.6%)
Total	215,333 (100%)	255,576 (100%)	288,084 (100%)	324,239 (100%)

The information offers a better understanding of flood extent and characteristics, such as water depths and flow velocities in flood-prone areas, which is the first step towards flood risk assessment and management. It is useful and essential information for stakeholders and authorities in contemplating management strategies and policies for flood damage reduction. Such policies and strategies include land use plans, floodplain building regulations, emergency preparedness and evacuation plans, as well as the development of structural mitigation measures to cope with flooding such as reservoirs and levee projects (Sanders, 2007; Sarhadi et al., 2012). For example, in the case of the Mun River Basin, floodplain management and regulations regarding agricultural, industrial and urban expansion should be developed and enforced, particularly at the upstream and the central parts of the basin, while levees are needed for protecting the exposed cities. The height of levees for different scenarios can be identified from the flood maps. This provides useful information for the decision-making process in relation to flood protection standards, which often depend on the economic status of the country. Also, an emergency plan explaining the activities necessary for dealing with adverse impacts from flooding should be prepared. The flood map can be used for education, communication, and in supplementary decision-making tools. Therefore, flood hazard maps should be easy to interpret and understand by various users (both technical and non-technical) and should be readily available to the public. A clear and precise understanding will increase public awareness and preparedness, which is a primary element of risk reduction (Jha, Bloch, & Lamond, 2012; Pilon, 2002).

The SWAT (including SWAT-CUP – a calibration program interface for SWAT) and HEC-RAS programs were adopted to develop the integrated model in this study because they are freely available, user-friendly, peer-reviewed are continuously improved and developed. The programs have extensive resources, including tutorials, examples, documentation, and platforms for resolving problems and bugs. They are widely accepted in many flood-related studies of which most employed either SWAT or HEC-RAS; however, in this study, both of them were integrated, which is rarely found in other studies. The capabilities and performances of the integrated model for flood modelling are satisfactory even in the present basin, which has a complex river system, and poor data quality and availability, as seen from their calibrated and validated results. This confirms the work of Loi et al. (2018), who developed an integrated model for real-time flood forecasting in Vietnam. Besides, HEC-DSS is a database system specially designed to efficiently store and retrieve large sets of data that are typically sequential (CEIWR-HEC, 2009). A good understanding of software features is critical because the program has its own structure and format. With the standard form, HEC-DSS facilitates the process of passing data from one analysis program to another.

The calibrated integrated model can also be used for the analysis and design of possible structural measures and alternatives (Popescu et al., 2010), or improved to establish a flood forecasting and warning system like in many countries (Gilles & Moore, 2010; Neal et al., 2011). This is the most effective tool to reduce the risk of loss of lives and economic damage from extreme rainfall (Loi et al., 2018; Pilon, 2002). The real-time inundation maps generated from forecasting systems are also an effective tool to inform relevant stakeholders, and can significantly assist in communication with residents in areas susceptible to flooding (Pilon, 2002).

It should be kept in mind that uncertainties exist in every stage of flood hazard mapping, from the beginning of the process (data collection, model selection, parameter selection, input data, model calibration, operation and handling of the models) until the outcome is obtained (Jha et al., 2012). The main limitations of this study are data quality and availability (e.g., missing rainfall and hydrologic data; unevenly distributed discharge and water level gauges with varying time series length and missing data; little field survey cross-section data and lack of hydraulic structure data along the river, such as bridges, weirs, etc.) contributing to uncertainties and inaccuracy in the results. The accuracy of the flood maps could be improved through the identification of possible sources of uncertainty and uncertainty analysis; however, it is beyond the scope of this study. We, therefore, recommended this for further research, as well as the integration of better data quality into the models if they are available.

4.5 Conclusion

In this study, an integrated hydrologic and hydraulic model, developed for the Mun River Basin, Thailand, is used to demonstrate the flood hazard mapping process for different probabilities. The models present forecasted runoff discharges, water levels, and water depths at 10, 25, 50, and 100-year return periods, which are necessary for deriving flood hazard maps. The maps include necessary and essential information according to international standards, which can help decision-makers to understand all aspects of flood characteristics, leading to appropriate flood mitigation measures, management strategies, and policies.

The contributions of the study are twofold. Firstly, the approach presented in this paper may be used as a prototype for flood hazard mapping in other basins in Thailand and elsewhere. In terms of model construction and validation, SWAT and HEC-RAS have excellent capabilities and functionality to serve hydrologic and hydraulic modeling purposes. Their calibration tools produce promising results even in the case of the Mun River, a complex river with low data quality. Better data quality and availability will enhance the accuracy of model prediction and reduce uncertainties. The validated models can be further advanced for the analysis and design of possible structural measures and flood forecasting and warning systems, which are important means of flood risk reduction. Additionally, the selected hydrologic parameters and their ranges from SWAT, and the calibrated Manning's n values can be used as a guide for model calibration in similar catchment characteristics and stream conditions. The latter is particularly beneficial in places where gauge data are not available.

The second major contribution is the provision of the first flood hazard maps at the Mun River Basin, Thailand, including the areas inundated at different severities. The upstream and the central parts of the river, and surrounding areas are more susceptible to flooding than the downstream environment, while deeper floods appear mostly in the downstream river floodplains. Rice fields are impacted the most because of their exposed locations. The flood maps help to identify potential hotspot areas where responses should be the priority, and more attention may be needed. This information is useful and essential for all relevant stakeholders to develop flood control measures, flood planning, and management to mitigate damage in the area. The maps must be regularly updated with all relevant data, i.e., input data, topography, land use, etc., to ensure that future decisions will be made based on the updated information. Furthermore, during the development of flood scenarios, the findings from rainfall analysis highlight the diversity of probability distributions, even at the basin scale. Thus the goodness of the assumed distribution should always be checked before performing design rainfall analysis. In general, we conclude that the use of integrated models to develop probabilistic flood hazard maps can be an important step in the future flood protection of the Mun River Basin and similar river systems in Thailand and abroad.

References

Abbaspour, K. C. (2015). SWAT-CUP: SWAT Calibration and Uncertainty Programs – A User Manual. *Eawag: Dübendorf, Switzerland*.

Abbaspour, K. C., Johnson, C., & Van Genuchten, M. T. (2004). Estimating uncertain flow and transport parameters using a sequential uncertainty fitting procedure. *Vadose Zone Journal, 3*(4), 1340-1352.

Abbaspour, K. C., Rouholahnejad, E., Vaghefi, S., Srinivasan, R., Yang, H., & Kløve, B. (2015). A continental-scale hydrology and water quality model for Europe: Calibration and uncertainty of a high-resolution large-scale SWAT model. *Journal of Hydrology, 524*, 733-752.

Alam, M. A., Emura, K., Farnham, C., & Yuan, J. (2018). Best-Fit Probability Distributions and Return Periods for Maximum Monthly Rainfall in Bangladesh. *Climate, 6*(1), 9.

Arnold, J. G., Kiniry, J. R., Srinivasan, R., Williams, J. R., Haney, E. B., & Neitsch, S. L. (2012). *Soil & Water Assessment Tool: Input/Output Documentation Version 2012.* Retrieved from

Arnold, J. G., Moriasi, D. N., Gassman, P. W., Abbaspour, K. C., White, M. J., Srinivasan, R., . . . Van Liew, M. W. (2012). SWAT: Model use, calibration, and validation. *Transactions of the ASABE, 55*(4), 1491-1508.

Blackburn, J., & Hicks, F. (2002). Combined flood routing and flood level forecasting. *Canadian Journal of Civil Engineering, 29*(1), 64-75.

Brunner, G. W. (2016a). *HEC-RAS River Analysis System Hydraulic Reference Manual Version 5.0.* Davis, CA, USA: US Army Corps of Engineers, Institute for Water Resources, Hydrologic Engineering Center (HEC).

Brunner, G. W. (2016b). *HEC-RAS River Analysis System, 2D Modeling User's Manual Version 5.0*. Davis, CA, USA: US Army Corps of Engineers, Institute for Water Resources, Hydrologic Engineering Center (HEC).

Brunner, G. W. (2016c). *HEC-RAS River Analysis System, User's Manual Version 5.0*. Davis, CA, USA: US Army Corps of Engineers, Institute for Water Resources, Hydrologic Engineering Center (HEC).

Bubeck, P., Kreibich, H., Penning-Rowsell, E. C., Botzen, W., De Moel, H., & Klijn, F. (2017). Explaining differences in flood management approaches in Europe and in the USA–a comparative analysis. *Journal of Flood Risk Management, 10*(4), 436-445.

Buishand, T. A. (1982). Some methods for testing the homogeneity of rainfall records. *Journal of Hydrology, 58*(1-2), 11-27.

CEIWR-HEC. (2009). *HEC-DSSVue, HEC Data Storage System Visual Utility Engine, User's Manual Version 2.0*. Davis, CA, USA: US Army Corps of Engineers, Institue for Water Resources, Hydrologic Engineering Center (HEC).

Centre for Research on the Epidemiology of Disaster (CRED). (2015). EM-DAT: The International Disaster Database. Retrieved January 21, 2015, from Centre for Research on the Epidemiology of Disaster - CRED, Université Catholique de Louvain, Brussels, Belgium www.emdat.be

Changnon, S. A. (2008). Assessment of flood losses in the United States. *Journal of Contemporary Water Research & Education, 138*(1), 38-44.

Chow, V. T. (1959). Open-channel hydraulics. In *Open-channel hydraulics*: McGraw-Hill.

Cooper, R. T. (2014). Open data flood mapping of Chao Phraya River basin and Bangkok Metropolitan Region. *British Journal of Environment and Climate Change, 4*(2), 186-216.

de Moel, H., van Alphen, J., & Aerts, J. C. J. H. (2009). Flood maps in Europe-methods, availability and use. *Nat. Hazards Earth Syst. Sci., 9*(2), 289-301. doi:https://doi.org/10.5194/nhess-9-289-2009

Demir, V., & Kisi, O. (2016). Flood hazard mapping by using geographic information system and hydraulic model: Mert River, Samsun, Turkey. *Advances in Meteorology, 2016*, 9. doi:http://dx.doi.org/10.1155/2016/4891015

Devi, G. K., Ganasri, B., & Dwarakish, G. (2015). A review on hydrological models. *Aquatic Procedia, 4*, 1001-1007.

Doong, D.-J., Lo, W., Vojinovic, Z., Lee, W.-L., & Lee, S.-P. (2016). Development of a new generation of flood inundation maps—A case study of the coastal city of Tainan, Taiwan. *Water, 8*(11), 521.

Douben, K.-J. (2006). Characteristics of river floods and flooding: a global overview, 1985–2003. *Irrigation and Drainage, 55*(S1), S9-S21. doi:10.1002/ird.239

European Parliament. (2007). Directive 2007/60/EC of the European Parliament and of the Council of 23 October 2007 on the assessment and management of flood risks.

Gilles, D., & Moore, M. (2010). *Review of Hydraulic Flood Modeling Software used in Belgium, The Netherlands, and The United Kingdom*. Retrieved from

Gupta, H. V., Sorooshian, S., & Yapo, P. O. (1999). Status of automatic calibration for hydrologic models: Comparison with multilevel expert calibration. *Journal of Hydrologic Engineering, 4*(2), 135-143.

IPCC. (2012). *Managing the Risks of Extreme Events and Disasters to Advance Climate Change Adaptation. A Special Report of Working Groups I and II of the Intergovernmental Panel on Climate Change*. Cambridge, UK, and New York, NY, USA: Cambridge University Press.

Jha, A. K., Bloch, R., & Lamond, J. (2012). *Cities and flooding: a guide to integrated urban flood risk management for the 21st century*: The World Bank.

Koivumäki, L., Alho, P., Lotsari, E., Käyhkö, J., Saari, A., & Hyyppä, H. (2010). Uncertainties in flood risk mapping: a case study on estimating building damages for a river flood in Finland. *Journal of Flood Risk Management, 3*(2), 166-183.

Krause, P., Boyle, D., & Bäse, F. (2005). Comparison of different efficiency criteria for hydrological model assessment. *Advances in Geosciences, 5*(5), 89-97.

Kukusamude, C., & Kongsri, S. (2018). Elemental and isotopic profiling of Thai jasmine rice (Khao Dawk Mali 105) for discrimination of geographical origins in Thung Kula Rong Hai area, Thailand. *Food Control, 91*, 357-364.

Kuntiyawichai, K. (2008). *Delineation of flood hazards and risk mapping in the Chi River Basin, Thailand*. Paper presented at the Proceedings of the 10th International Drainage Workshop of ICID Working Group on Drainage, Helsinki University of Technology.

Kuntiyawichai, K. (2012). *Interactions between Land Use and Flood Management in the Chi River Basin*. (Ph.D. dissertation), UNESCO-IHE Institute of Water Education and Wageningen University, Delft, the Netherlands.

Kwak, Y., Park, J., Yorozuya, A., & Fukami, K. (2012). *Estimation of flood volume in Chao Phraya River basin, Thailand, from MODIS images coupled with flood inundation level*. Paper presented at the 2012 IEEE International Geoscience and Remote Sensing Symposium.

Laat, P. J. M. d. (2012). *Workshop on Hydrology*: UNESCO-IHE Institute for Water Education.

Loi, N. K., Liem, N. D., Tu, L. H., Hong, N. T., Truong, C. D., Tram, V. N. Q., . . . Jeong, J. (2018). Automated procedure of real-time flood forecasting in Vu Gia–Thu Bon river basin, Vietnam by integrating SWAT and HEC–RAS models. *Journal of Water and Climate Change*, 11. doi:https://doi.org/10.2166/wcc.2018.015

Mekong River Commission Secretariat. (2014). *Training Manual: ArcSWAT 2012*: Technical Support Division, Mekong River Commission Secretariat.

Moreda, F., Gutierrez, A., Reed, S., & Aschwanden, C. (2009). *Transitioning NWS operational hydraulics models from FLDWAV to HEC-RAS*. Paper presented at the World Environmental and Water Resources Congress 2009: Great Rivers.

Moriasi, D. N., Arnold, J. G., Van Liew, M. W., Bingner, R. L., Harmel, R. D., & Veith, T. L. (2007). Model evaluation guidelines for systematic quantification of accuracy in watershed simulations. *Transactions of the ASABE, 50*(3), 885-900.

Mosquera-Machado, S., & Ahmad, S. (2007). Flood hazard assessment of Atrato River in Colombia. *Water Resources Management, 21*(3), 591-609.

Motovilov, Y. G., Gottschalk, L., Engeland, K., & Rodhe, A. (1999). Validation of a distributed hydrological model against spatial observations. *Agricultural and Forest Meteorology, 98*, 257-277.

Nakmuenwai, P., Yamazaki, F., & Liu, W. (2017). Automated extraction of inundated areas from multi-temporal dual-polarization RADARSAT-2 images of the 2011 central Thailand flood. *Remote Sensing, 9*(1), 78.

Nash, J. E., & Sutcliffe, J. V. (1970). River flow forecasting through conceptual models part I — A discussion of principles. *Journal of Hydrology, 10*(3), 282-290.

Neal, J., Schumann, G., Fewtrell, T., Budimir, M., Bates, P., & Mason, D. (2011). Evaluating a new LISFLOOD-FP formulation with data from the summer 2007 floods in Tewkesbury, UK. *Journal of Flood Risk Management, 4*(2), 88-95.

Phoonphongphiphat, A. (Producer). (2018, 31 May 2019). Thailand diversifies rice production to retain global market share. *Nikkei Asian Review*. Retrieved from https://asia.nikkei.com/Business/Markets/Commodities/Thailand-diversifies-rice-production-to-retain-global-market-share

Pilon, P. J. (2002). *Guidelines for reducing flood losses*. Retrieved from https://www.unisdr.org/we/inform/publications/558

Popescu, I., Jonoski, A., Van Andel, S., Onyari, E., & Moya Quiroga, V. (2010). Integrated modelling for flood risk mitigation in Romania: case study of the Timis–Bega river basin. *International journal of river basin management, 8*(3-4), 269-280.

Raes, D., Willems, P., & Gbaguidi, F. (2006). *RAINBOW–A software package for hydrometeorological frequency analysis and testing the homogeneity of historical data sets*. Paper presented at the Proceedings of the 4th International Workshop on Sustainable management of marginal drylands. Islamabad, Pakistan.

Rakwatin, P., Sansena, T., Marjang, N., & Rungsipanich, A. (2013). Using multi-temporal remote-sensing data to estimate 2011 flood area and volume over Chao Phraya River basin, Thailand. *Remote sensing letters, 4*(3), 243-250.

Royal Irrigation Department of Thailand. (2018). Reservoir data. from the Royal Irrigation Department of Thailand http://app.rid.go.th:88/reservoir/

Sanders, B. F. (2007). Evaluation of on-line DEMs for flood inundation modeling. *Advances in Water Resources, 30*(8), 1831-1843.

Sarhadi, A., Soltani, S., & Modarres, R. (2012). Probabilistic flood inundation mapping of ungauged rivers: Linking GIS techniques and frequency analysis. *Journal of Hydrology, 458*, 68-86.

Son, N., Chen, C., Chen, C., & Chang, L. (2013). Satellite-based investigation of flood-affected rice cultivation areas in Chao Phraya River Delta, Thailand. *ISPRS journal of photogrammetry and remote sensing, 86*, 77-88.

Tingsanchali, T., & Karim, F. (2010). Flood-hazard assessment and risk-based zoning of a tropical flood plain: case study of the Yom River, Thailand. *Hydrological Sciences Journal–Journal des Sciences Hydrologiques, 55*(2), 145-161.

Van Alphen, J., & Passchier, R. (2007). Atlas of Flood Maps, examples from 19 European countries, USA and Japan, Ministry of Transport. *Public Works and Water Management, The Hague, Netherlands, prepared for EXCIMAP, available at: http://ec. europa. eu/environment/water/flood risk/flood atlas/index. htm.*

Van Liew, M. W., Veith, T. L., Bosch, D. D., & Arnold, J. G. (2007). Suitability of SWAT for the conservation effects assessment project: Comparison on USDA agricultural research service watersheds. *Journal of Hydrologic Engineering, 12*(2), 173-189.

Vojinovic, Z., Hammond, M., Golub, D., Hirunsalee, S., Weesakul, S., Meesuk, V., . . . Abbott, M. (2016). Holistic approach to flood risk assessment in areas with cultural heritage: a practical application in Ayutthaya, Thailand. *Natural Hazards, 81*(1), 589-616.

W. Van Liew, M., G. Arnold, J., & D. Garbrecht, J. (2003). Hydrologic Simulation on Agricultural Watersheds: Choosing between two models. *Transactions of the ASAE, 46*(6), 1539-1551. doi:https://doi.org/10.13031/2013.15643

Winchell, M., Srinivasan, R., Luzio, M. D., & Arnold, J. (2013). *ArcSWAT Interface For SWAT2012, User's Guide* Temple, Texas, USA.

5

Drought hazard assessment

This chapter is published as:
Prabnakorn, S., Maskey, S., Suryadi, F., & de Fraiture, C. (2019). Assessment of drought hazard, exposure, vulnerability, and risk for rice cultivation in the Mun River Basin in Thailand. *Natural Hazards*, 97, 891-911. doi:10.1007/s11069-019-03681-6.

Abstract

When assessing drought risk, most studies focus on hazard and vulnerability, paying less attention to exposure. Here, we propose a comprehensive drought risk assessment scheme combining hazard, exposure, and vulnerability. At the Mun River Basin, 90% of rice cultivation is rain-fed and regularly encounters droughts resulting in the lowest yields in the country. The water deficit calculated with respect to rice water requirement is used to assess drought hazard and is estimated at monthly time steps. We use drought severity and frequency for hazard estimation and population and rice field characteristics for exposure. Vulnerability is represented by physical and socioeconomic factors and coping and adaptive capacity. Between 1984 and 2016, monthly precipitation during the rice-growing season was insufficient to meet rice water needs at all growth stages (July - November). The hazard is more severe in October and November, which can lead to significantly reduced yields. People and rice fields in the center part of the basin are more exposed to droughts than in other parts. Extensive areas are under high and moderate vulnerability due to low coping capacity. The higher drought risks appear in the last two months of the growing season and decrease from north to south, while the risk map of total precipitation demonstrates that most of the areas have low and very low risk. This emphasizes the importance of monthly time series analysis to calculate agricultural drought hazard and risk. Consequently, we recommend using the hazard and risk maps for October and November instead of the total precipitation to develop solutions to improve rice yield.

Keywords: Precipitation, Monthly time series analysis, Risk assessment scheme, Water deficit, Agricultural drought

5.1 Introduction

Drought is a frequently occurring and widespread natural hazard, which adversely affects many people and economic sectors among which agriculture is most affected (FAO, 2017, 2018; World Meteorological Organization (WMO) and Global Water Partnership (GWP), 2016). Drought creates unfavourable conditions for crop development and eventually reduces yields. It occurs not only in arid regions but also in regions with relatively abundant precipitation (Pereira, Cordery, & Iacovides, 2002). Thailand, for example, has an average precipitation of about 1455 mm/year; however, it periodically endures droughts, and agriculture is the main sector affected. In 2013 about 3850 km2 of agricultural areas and 9 million people were affected by drought with the total damage of 90 million US$. The damage in 2014 was even greater (550 million US$) with 6800 km2 and 9 million people affected (Department of Water Resources of Thailand, 2016). The level of impacts depends on socioeconomic conditions, such as population density, types of crops, poverty, an education level (World Meteorological Organization (WMO) and Global Water Partnership (GWP), 2016). Therefore, understanding the hazard extent, exposure, vulnerability, and risk of droughts to agriculture is essential for developing appropriate water management and mitigation measures that can reduce the adverse impacts and enhance crop yields.

Different concepts and definitions to drought risk assessment have been introduced and adopted. For example, risk is defined as the product of the hazard and vulnerability stated by Knutson, Hayes, and Phillips (1998); and Wilhite (2000), and was adopted by He et al. (2013); Kim, Park, Yoo, and Kim (2015); Shahid and Behrawan (2008) in Bangladesh, China and South Korea respectively. Risk is a product of the exceeding probability of drought hazard at a specific severity, and its relevant drought disaster affected ratio by Lei and Luo (2011) in China. Risk in Germany is examined with respect to water supply, which calculates from precipitation, plant-available water and recharges from groundwater, and yield potential for cereals, root crops and grass (Schindler, Steidl, Müller, Eulenstein, & Thiere, 2007). An agricultural drought risk assessment model specific to corn and soybean in Nebraska, USA was developed based on the standardized precipitation index (SPI) and crop-specific drought index (CSDI) by using multivariate techniques (H. Wu, Hubbard, & Wilhite, 2004; H. Wu & Wilhite, 2004). Risk in northeast Thailand was assessed using remote sensed data and Geographic Information Systems (GIS). The analysis derived meteorological, hydrological, and physical droughts and finally integrated them by matrix analysis (Mongkolsawat et al., 2001). However, when evaluating drought risk, most studies focus on hazard and vulnerability (Belal, El-Ramady, Mohamed, & Saleh, 2014). Some consider exposure as part of vulnerability, but the distinction between vulnerability and exposure is often not made explicit (O.D. Cardona et al., 2012; Liu, Wang, Peng,

Braimoh, & Yin, 2013). Less attention is devoted to understanding and addressing exposure.

Drought indices, especially the Standardized Precipitation Index (SPI) and Standardized Precipitation Evapotranspiration Index (SPEI), have been widely adopted in drought hazard analysis at all scales: global (Spinoni, Naumann, Carrao, Barbosa, & Vogt, 2014; Wang et al., 2014), regional (Blauhut, Gudmundsson, & Stahl, 2015; Jeong, Sushama, & Naveed Khaliq, 2014; Keyantash & Dracup, 2004), national (Daneshvar, Bagherzadeh, & Khosravi, 2013; He, Lü, Wu, Zhao, & Liu, 2011; Kim et al., 2015) and basin (S. Pandey, Pandey, Nathawat, Kumar, & Mahanti, 2012; Tsakiris, Pangalou, & Vangelis, 2007). Although the SPEI has a strong relationship with rice yield (Prabnakorn, Maskey, Suryadi, & de Fraiture, 2016), it is not clear whether standardized indices like the SPEI or SPI can identify areas that are more prone to droughts than others (Lloyd-Hughes & Saunders, 2002). In addition, some studies conducted drought hazard analysis by linking the climatic variables to rice production, such as the research by Chung, Jintrawet, and Promburom (2015); Li, Angeles, Radanielson, Marcaida, and Manalo (2015); Prabnakorn, Maskey, Suryadi, and de Fraiture (2018); Rahman, Kang, Nagabhatla, and Macnee (2017); Sarker, Alam, and Gow (2012) from which their concepts significantly contribute to the method using to estimate the drought hazard index (DHI) for assessing drought hazard in this study.

Drought is a complex phenomenon composed of natural, physical, and social components (Hayes, Wilhelmi, & Knutson, 2004; Wilhite, 2000). A clear definition is needed to avoid confusion about the onset of drought and its severity (Hayes et al., 2004; Yevjevich, 1967). In this study, we follow the agronomist's view that drought is a condition of water stress and shortage, which affects crop growth and reduces yields (Maracchi, 2000; Pereira et al., 2002). Hence, in this study, the risk was assessed by using drought hazard, exposure, and vulnerability, associated with rice growth and yields.

This paper aims to characterize spatial variations in drought hazard, exposure, vulnerability, and risk for rice cultivation. We propose a new scheme for drought risk assessment adapted from Hagenlocher, Renaud, Haas, and Sebesvari (2018). The concept is based on the three key determinants of drought risk, i.e., hazard, exposure, and vulnerability, with each component contributing to the composite risk in the same degree. We use the percentage of water deficit that affects rice production to derive the DHI, which to our knowledge has not been done before. The method concerns both drought severity and frequency, the frequency of each severity class is calculated and used to derive drought hazard directly, which is more sensitive to changes in drought severity than the previous method (He et al., 2011; He et al., 2013; Kim et al., 2015;

Shahid & Behrawan, 2008). We assess drought hazard and risk at monthly time steps rather than only using the total precipitation over the growing season. This is important because the sensitivity of crops to drought is different at different growth stages. To understand drought risk in all dimensions, we include population and rice field characteristics in the estimation of exposure, and physical and socioeconomic factors and coping and adaptive capacity in vulnerability analysis.

5.2 Materials and methods

5.2.1 Constructing a drought risk assessment framework

Concepts and quantitative approaches to risk assessment are continuously evolving. A widely accepted holistic concept integrates and links all dimensions associated to risk such as physical/natural, social, economic, political and environmental aspects (Hagenlocher et al., 2018; IPCC, 2012, 2014; Sebesvari et al., 2016). Risk is a function of these factors that are usually
classified into three key components: hazard, exposure, and vulnerability (Omar D Cardona, 2011; IPCC, 2012, 2014; Koks, Jongman, Husby, & Botzen, 2015; Kron, 2005). The simple mathematical form of risk is the product of these three components (5.1) (Carrão, Naumann, & Barbosa, 2016; Kron, 2002; Peduzzi, Dao, Herold, & Mouton, 2009).

$$Risk = Hazard \times Exposure \times Vulnerability, \qquad (5.1)$$

Consistent with this concept, we developed a framework of drought risk assessment presented in Figure 5-1. Hazard analysis takes into account drought severity and the probability of occurrence in the past 33 years (1984-2016). The exposure and vulnerability include multiple components which are described in the remainder of the paper. Consequently, Eq.5.1 is modified to

$$DRI = DHI \; x \; DEI \; x \; DVI, \qquad (5.2)$$

where DRI is drought risk index, DHI is drought hazard index, DEI is drought exposure index and DVI is drought vulnerability index. Owing to the product relationship, the likelihood of drought risk equals zero if there is no hazard or no exposure at a given location (or if neither is present).

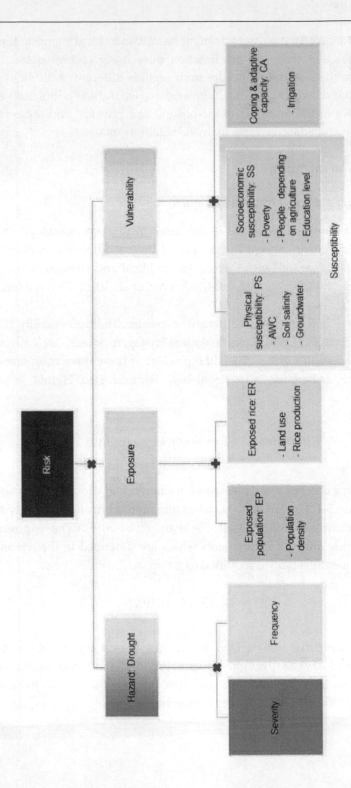

Figure 5-1 Drought risk assessment scheme concept for a single hazard adapted from Hagenlocher et al. (2018).

5.2.2 Estimation of drought hazard

Continuous time-series data of monthly precipitation over the period of 1984 to 2016 were collected from 53 selected stations distributed over the basin (Figure 2-2). The data was verified and validated as follows. A cross-database comparison was performed between precipitation data from two sources: the Royal Irrigation Department and the Meteorological Department of Thailand. The values that were flagged as outliers, incorrect or doubtful were then compared with the values from neighboring stations and were corrected. The missing precipitation values were replaced, where possible, by estimated values using multiple linear regression techniques. The method is based on fitting the best straight line through neighboring stations (independent variables). The set of neighboring stations giving the highest coefficient of determination (ρ^2) were selected for data completion, but the missing values were retained in the record if $\rho^2 < 0.5$ to ensure data quality and reliability.

We assess drought hazard by evaluating an imbalance between water supply and demand. Drought occurs when the demand is higher than the supply. Water supply is calculated as the monthly or total precipitation over the rice-growing seasons. Water demand refers to the water required for rice growth. The calculation is based on Brouwer, Prins, and Heibloem (1989), which includes the water needs for soil saturation, evapotranspiration, percolation and seepage losses, and the establishment of a water layer. The equation takes the form

$$WR_i = SAT + ET_{rice} + PERC + WL, \tag{5.3}$$

where WR_i is water requirement for rice at month i or the total water requirement for the entire growing season (mm). SAT is the amount of water needed to saturate the soil (mm) at the beginning of the growing season, which depends on the soil type and rooting depth. It is the area-weighted average of the available water contents of all soils in the study area at 1 m effective rooting depth of rice (Allen, Pereira, Raes, & Smith, 1998). The rice water needs $ET_{rice} = ET_0 \times K_c$; the ET_0 is the reference evapotranspiration rate estimated by using the ET_0 calculator software provided by FAO (2009), and the K_c is the crop factor for rice obtained from . PERC is the percolation and seepage losses, which depend on the soil type, and WL is the amount of water needed for the establishment of a water layer (mm); both values obtained from Brouwer et al. (1989).

Then the percentage of water deficit (WD_i) can be calculated using

$$WD_i = \begin{cases} \dfrac{WR_i - P_i}{WR_i} & \text{if } WR_i > P_i \\ 0 & \text{if } WR_i \leq P_i \end{cases} \qquad (5.4)$$

where P_i is monthly or total precipitation (mm/month or mm/growing season), the WD_i values were categorized into five levels and assigned weights ranging between 0 and 4 according to their severity (Table 5-1).

Table 5-1 Weights for different levels of water deficit

% Water Deficit (WD)	Weight (W_j)
No water deficit	0
0-25	1
25-50	2
50-75	3
75-100	4

The frequency of each level of WD_i is calculated using

$$F_j = \frac{n_j}{N}, \qquad (5.5)$$

where F_j is drought frequency at severity level j, n_j is the number of years with water deficit at level j, and N is the total number of years under consideration.

Finally, the drought hazard index is calculated by integrating the product of the weight of water deficit at each level (W_j) and its relevant frequency (F_j). The minimum value of DHI is 0, meaning there is enough rainwater for rice cultivation every year (no water deficit). The maximum amount is $4 = 4 \times 1$, meaning there is insufficient water for rice every year. The equation takes the form

$$DHI = \sum_{j=0}^{4} (W_j \times F_j), \qquad (5.6)$$

Figure 5-2 depicts the DHI values derived by this method (triangle), showing it is more sensitive to changes of drought severity than the previous method (circle) used by He et al. (2011); He et al. (2013); Kim et al. (2015); Shahid and Behrawan (2008). The method assigned weights to each level of drought severity and also assigned ratings to each range of the frequency; the DHI is the summation of the product of the weight and its relevant rating. The double rating steps (to both severity and frequency) thus reduce the sensitivity of the DHI.

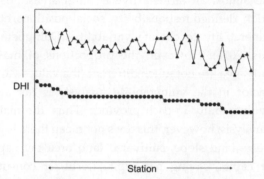

DHI

Station

Figure 5-2 Comparison of October DHI values at all stations obtained from the new method used in this study (triangle) and the previous method (circle).

All DHI values were used to derive drought hazard maps using the ordinary kriging method in Geostatistical Analyst in ArcGIS. The method has been shown to outperform universal kriging slightly and is significantly superior to the inverse distance weighting methods over various types of surfaces and all levels of noise and spatial correlation (Zimmerman, Pavlik, Ruggles, & Armstrong, 1999). Finally, the hazard maps are divided into 4 levels based on the DHI values: very low, low, moderate, and high.

5.2.3 Indicator selection for drought exposure and vulnerability

Following the drought risk assessment concept in Figure 5-1, we selected population density and rice field location (derived from land use maps) and production as the most visible elements to represent exposure. The vulnerability analysis can include many factors, depending on the available data. Hagenlocher et al. (2018) provided a library of indicators, which can be used as a blueprint for vulnerability assessment of deltaic environments, as guidance for other environments. In our case where data availability is a constraint, we selected the factors based on the literature review, focusing on studies with limited data. Details on the literature review and results are presented in Table A-5.1 in the Appendix.

As our study focuses on agricultural drought in rice cultivation, we decided to use the factors most commonly used in agricultural drought vulnerability assessment: these are soil water-holding capacity, poverty rate, the proportion of people depending on agriculture, an education level (average year of schooling), and irrigation support. Furthermore, we included soil salinity and groundwater quantity and quality in our study, although these have rarely been used because they are significant for crop growth and production. In contrast, the proportion of people living in a rural area is disregarded because almost all farmers live in rural areas. Also owing to gender inequality in education, defined responsibility, social positions, etc., men and women reveal different vulnerability to climate variability (Dah-gbeto & Villamor, 2016; Djoudi & Brockhaus, 2011). However, the proportions of male to female in the provinces in our study area are not much different (the values vary from 0.97-1.0), the inclusion of this factor in the vulnerability analysis is not significantly made a difference to total vulnerability in each province. Thus, the male to female ratio is neglected from the analysis; however, this does not mean there is gender equality. The remaining factors (e.g., land slope, cultivated land area per capita, gross domestic product (GDP) per capita, saving per capita, fertilizer consumption, agriculture machinery, and institutional capacity) were disregarded, mainly because little data was available; however, we do not believe that this would have changed our findings significantly, and these have seldom been included in the previous studies as detailed in Table A-5.1 in the Appendix.

5.2.4 Estimation of drought exposure and vulnerability

According to IPCC (2012, 2014), exposure is the presence of people, society, livelihoods, ecosystem, environment, resources, infrastructure or economic or cultural assets that could be adversely affected by the hazards. Vulnerability is the propensity or predisposition to be adversely affected; it is determined by a combination of social, economic, environmental, technological, and physical factors (Omar D Cardona, 2011; Hayes et al., 2004; Shahid & Behrawan, 2008). Following the modular scheme of drought risk assessment, the calculation within each exposure (EP and ER) and vulnerability (PS, SS, and CA) units were achieved by using the following equation:

$$EP, ER, PS, SS, and\ CA = \sum_{i=1}^{n} (a_i \times X_i), \qquad (5.7)$$

where X_i is exposure or vulnerability factors, and α_i is the weights given to factors. This weighted average formulation is also applied to estimate DEI for exposure and DVI for vulnerability. There are various techniques for assigning weights, as mentioned above. Here we used equal weights to all factors to finally obtain DEI (Eq. 5.8) and DVI (Eq. 5.9).

$$DEI = \frac{EP + ER}{2} \qquad (5.8)$$

$$DVI = \frac{PS + SS + CA}{3} \qquad (5.9)$$

Lastly, the results of all raster cells were obtained using the Spatial Analyst tool in ArcGIS. The results were classified into four levels using natural breaks: very low, low, moderate, and high. The natural breaks represent natural groupings inherent in the data. This is an iterative process to find the proper intervals with the variance within the same class as small as possible and that between classes maximized (De Smith, Goodchild, & Longley, 2009).

5.2.5 Data collection, processing, and weighting

An overview of the data used and their sources are presented in Table 5-2. Land use and irrigation support data are in gridded format and can directly be used for the next weighting step. However, some gridded data, i.e., soil type, soil salinity and groundwater quantity (expected well yield) and quality (total dissolved solids: TDS), and non-gridded format data, i.e., rice production, population density, poverty rate, education level (average year of schooling), as a result of a paucity of spatially explicit data, need some data processing and preparation. For the soil data, we derived the available water content at a specific rooting depth, which is constant in a particular soil type but varies widely between soil texture and structure (Brouwer, Goffeau, & Heibloem, 1985; Wilhelmi & Wilhite, 2002). It is the difference in water content between field capacity and permanent wilting point, and in this case, the effective rooting depth of rice (in the absence of characteristics that can restrict rooting depth, such as bedrock, lithologic discontinuities, water tables) is 1 m. Soil salinity, and groundwater quantity (expected well yield) and quality (TDS) data were categorized into different levels by the experts from responsible departments. The non-gridded data were collected and arranged at administrative units, i.e., district and province, before transforming into the gridded format by using ArcGIS.

Table 5-2 List of data used in drought risk analysis (with year and source)

Factor	Data	Gridded dataset	Year	Source
Precipitation	Monthly precipitation		1984-2016	Royal Irrigation Department Meteorological Department
Population density	Population density		2016	Department of Provincial Administration
Land use	Land use	√	2013	Land Development Department
Rice production	Rice production by district		2016	Office of Agricultural Economics
Soil	Soil type map	√	2009-2010	Land Development Department
Saline soil	Soil salinity map	√	2006	Land Development Department

Factor	Data	Gridded dataset	Year	Source
Groundwater quantity and quality	Expected well yield and TDS	√	2011-2012	Department of Groundwater Resources
Poverty rate	Poverty rate		2013	National Statistical Office
People depending on agriculture	People depending on agriculture		2016	Office of Agricultural Economics
Education level	Average year of schooling		2016	Office of the Education Council
Irrigation support	Irrigation	√	2012-2013	Royal Irrigation Department

Assigning weights to the different factors and indicators are the key to developing a composite vulnerability index. There is no generally accepted weighting for factors and indicators: Wilhelmi and Wilhite (2002) assigned the weights based on the relative contribution of each factor to overall vulnerability, Chen, Cutter, Emrich, and Shi (2013) applied a statistic-based method (Principal Component Analysis: PCA) to evaluate the weights of the factors in social vulnerability assessment, and Yuan et al. (2015) assigned the weights based on expert opinions and employed accelerating genetic algorithm-based analytic hierarchy process (AGA-AHP) to quantify the opinion. In our study, as described above when we obtained the data, some data were already classified into different classes, thus we followed Wilhelmi and Wilhite (2002), because their method is widely accepted in vulnerability assessment where factors contain several classes (S. Pandey et al., 2012; Safavi, Esfahani, & Zamani, 2014; Shahid & Behrawan, 2008; J. Wu et al., 2011). Consequently, each class of all factors was given a weight between 0 and 4 (0 represents non-affected, 1 represents least significant and 4 represents the most significant). For example, soil types with less than 150 mm water-holding capacity are ranked 4, showing that they are less able to withstand a deficiency of precipitation than ones with higher water-holding capacity. Dummy weights were assigned to some classes for masking purposes, and they were excluded from the analysis (Table 5-3 and Table 5-4).

Table 5-3 Weighting scheme for drought exposure analysis

Factor		Class	Weight
Exposed population	Population density: PD (people/km^2)	<=100	1
		100-150	2
		150-200	3
		> 200	4
Exposed rice	Land use type: LU	Paddy rice field	4
		Others	50 (masking)
	Rice production: RP (ton/ha)	≤ 2	1
		2 – 2.33	2
		2.33 – 2.66	3
		> 2.66	4

Table 5-4 Weighting scheme for drought vulnerability analysis

	Factor	Class	Weight
Physical susceptibility	Soil water-holding capacity: SM (mm)	≤ 150	4
		150-200	3
		200-250	2
		> 250	1
		Unidentified: wetland, urban, etc.	20 (masking)
	Saline soil level: SL (% of salt crusts on the surface)	High salt (10-50%)	4
		Moderate salt (1-10%)	3
		Slight salt (<1%)	2
		Potential - Bedded rock salt	1
		Non-affected	0
		Unidentified: forest, water, etc.	30 (masking)
	Groundwater quantity and quality: GW	Yield < 5 m³/hr., TDS > 1500 mg/l	4
		Yield < 2 m³/hr., TDS 500-1500 mg/l	3.4
		Yield 5-10 m³/hr., TDS > 1500 mg/l	
		Yield < 2 m³/hr., TDS < 500 mg/l	2.8
		Yield 2-10 m³/hr., TDS 500-1500 mg/l	
		Yield 10-20 m³/hr., TDS > 1500 mg/l	
		Yield 2-10 m³/hr., TDS < 500 mg/l	2.2
		Yield 10-20 m³/hr., TDS 500-1500 mg/l	
		Yield > 20 m³/hr., TDS > 1500 mg/l	
		Yield 10-20 m³/hr., TDS < 500 mg/l	1.6
		Yield > 20 m³/hr., TDS 500-1500 mg/l	
		Yield > 20 m³/hr., TDS < 500 mg/l	1
		Surface water	40 (masking)
Social susceptibility	Poverty: PV (%)	<=10	1
		10-20	2
		20-30	3
		30-40	4
	People depending on agriculture: PA (%)	< 40	1
		40-50	2
		50-60	3
		> 60	4
	Education level – average year of schooling: ED (year)	> 12	1
		9-12	2
		6-9	3
		< 6	4
Coping & adaptive capacity	Irrigation support: IS	Available	0
		Not available	4

We investigated statistical correlations between the factors in exposure and vulnerability indices to verify whether a multicollinearity problem exists by using Kendall's Tau correlation coefficient (T_b). This measures associations between ranked data and the results range from -1 and 1. The correlation values were then examined regarding statistical significance through a two-tailed approach. If $T_b > 0.9$ and statically significant, the data are highly correlated (Hagenlocher et al., 2018). The analysis was carried out in IBM SPSS Statistics, and the results are in Table A- 5.2, Table A- 5.3, Table A- 5.4 in the Appendix, no issue of multicollinearity was detected. Consequently, all exposure and vulnerability factors can be used in the study.

5.3 Results and Discussions

5.3.1 Hazard

The average, minimum, and maximum monthly precipitation during the rice-growing season from 1984-2016 are shown in Figure 5-3 for all 10 provinces in the basin. September has the highest amount of average precipitation for all except the three eastern provinces, Yasothon, Amnat Charoen and Ubon Ratchathani, where the highest average precipitation is in August. Additionally, the precipitation decreases from east to west in the rainy months, July, August, and September. This is because the provinces to the west are on the leeward side of the mountain ranges against the southwest monsoon (Nawata, Nagata, Sasaki, Iwama, & Sakuratani, 2005). Moreover, those provinces are further inland, and they receive less of the cyclonic rains that occur annually in the South China Sea and blow westward to Thailand. The precipitation declines in October and reaches a low in November; the amounts in these two months do not differ much among provinces. The variations in precipitation (max-min) are high in the rainy months, July, August, and September, and low in the dry months, October and November. The variations in precipitation in July and August are higher in the eastern provinces, Si Sa Ket, Yasothon, Amnat Charoen and Ubon Ratchathani, and the variations in September are higher in Maha Sarakham and the eastern provinces.

Table 5-5 illustrates the water requirement for rice with a growth period of 120 days (mid-July to mid-November). Over the growing season, a total of 1351 mm water is needed; the monthly requirement is high at the beginning of the growing season when land is prepared, and in October when the rice needs the highest water layer (up to 100 mm). The water is lower in August and September (20-50 mm), and no standing water is required in November, resulting in lower water requirements.

Over the past 33 years, many years have had insufficient monthly precipitation during the growing season to meet the water requirements for rice at all phases (Figure 5-4). If we only consider the total precipitation over the growing season, it is not far below the amount required for rice growth. The hazard map of the growing season shows only moderate and low levels of drought hazard; the area of moderate hazard in the west accounts for 57% of the total area, while the rest (43%) in the east has a low hazard.

November represents the most severe water shortage in terms of intensity and frequency. However, the dry conditions in this month will not cause an adverse effect on production since this month represents the final ripening phase; instead, they homogenize maturation and facilitate harvesting (Wopereis, Defoer, Idinoba, Diack, & Dugué, 2008).

Rice is somewhat fragile in October (mid-reproductive to mid-ripening phases) as two crucial stages occur at this time: flowering and grain filling. Although flowering occurs around mid-October, it is essential to maintain the water layer throughout the month because yield will decrease considerably if standing water disappears before flowering, as this causes a high percentage of spikelet sterility (Bouman, Lampayan, & Tuong, 2007; Jearakongman et al., 1995). The model simulation indicates that KDML105 yield falls below 1 ton/ha when the water layer disappears more than 20 days before flowering (Fukai, Basnayake, & Cooper, 2000). The grain-filling stage occurs in the second half of the month when the rice accumulates carbohydrates through photosynthesis. This stage determines the individual grain weight and quality, and it is susceptible to water deficiency, which causes incomplete grain filling (Beighley, 2010; Yoshida, 1981). Therefore, rice fields in the upper part of the basin (48% of the total area), which face a high level of drought, are more vulnerable to yield reduction than those in the lower part, which only suffer a moderate level of drought.

September has the highest precipitation and the least water shortage; 20% of the eastern area is at very low risk, and the rest (80%) is at low risk. In this month, panicle initiation and differentiation are the critical stages that influence the yield because the number of spikelets per panicle is determined at this stage. The initiation and development of panicle primordia are mainly controlled by short photoperiods (Yoshida, 1981) because both KDML105 and RD6 are photoperiod-sensitive (Bureau of Rice Research and Development (BRRD), n.d.). Thus, if the days are too long than critical photoperiod, those stages will be delayed, and the length of the growing period will be extended, which may increase the risk of drought at the late stage when the rainy monsoon season has ended.

In August and July, rice in the western region is more vulnerable to drought than that in the east. Drought in August, before or during tillering development, decreases the number of tillers and panicles per unit area (Bouman et al., 2007) which in turn leads to reduced yields. In July, the rainwater is mainly used for land preparation; thus, a dry spell will postpone the onset of the growing season, which increases the risk at the late stage.

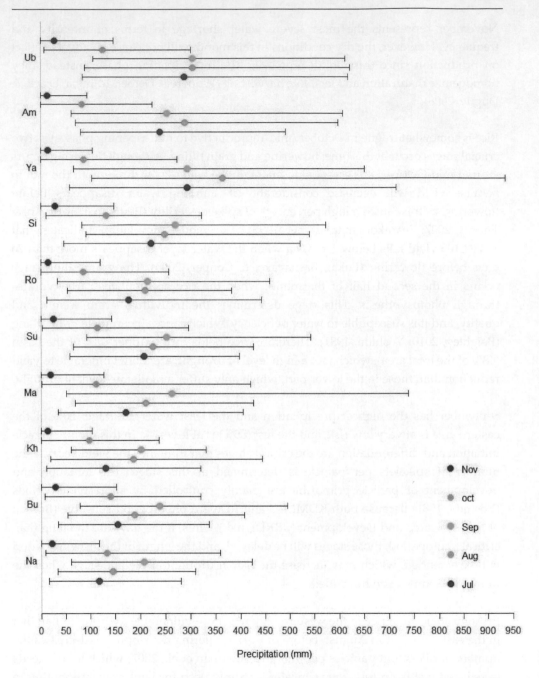

Figure 5-3 Average (dot), minimum (lower bar) and maximum (upper bar) monthly precipitations during the rice growing period from 1984-2016. The provinces from top to bottom represent from east to west of the basin.

Table 5-5 Total water requirement for rice cultivation with a life cycle of 120 days

Growing season	Jul	Aug	Sep	Oct	Nov	Total
Growth stage		Vegetativ	Reproducti	Ripenin		
Total water requirement (mm)	323	296	287	333	113	1351

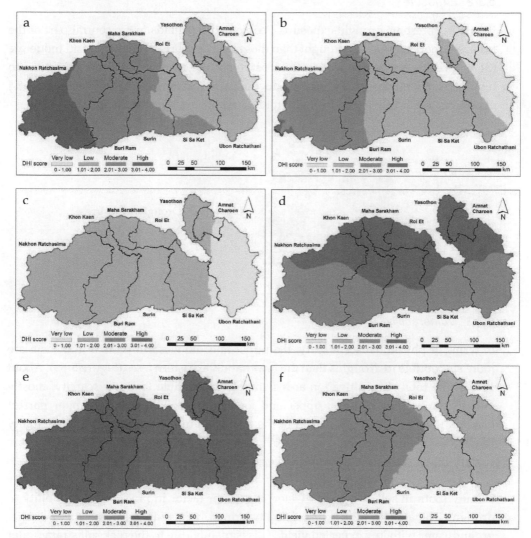

Figure 5-4 Hazard maps of months Jul (a), Aug (b), Sep (c), Oct (d), Nov (e) and over the entire rice-growing season (f).

Drought hazards in July, August, September and over the growing season decrease from west to east following the same pattern as precipitation distribution (in the opposite direction), whereas drought hazard in October decreases from north to south. This may be because there is less influence from the southwest monsoon. Moreover, in October and November, the tropical cyclones rarely move to the lower part of the northeast, where the basin is situated, but usually, move to other regions of the country instead.

5.3.2 Exposure

Figure 5-5 presents drought exposure. People and rice fields in the central part of the basin are more exposed to drought than those in other parts. High (27%) and moderate (31%) levels of exposures are mostly observed at the central part of the basin due to high population density, and rice production, whereas low (36%) and very low (6%) exposures are mostly found in Nakhon Ratchasima and the eastern region.

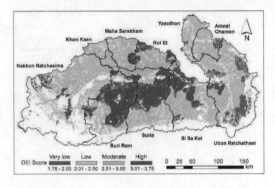

Figure 5-5   Drought exposure

5.3.3 Vulnerability

In terms of physical susceptibility, the upper and the western parts of the basin are more susceptible to droughts than the lower and the eastern ones (Figure 5-6a). About 13% of the total rice cultivation area is highly susceptible to water scarcity, mostly around the borders between upper and lower provinces and in the upper part of Nakhon Ratchasima. 34% and 41% of the area have moderate and low susceptibility respectively, with patches scattered over the basin. 12% of the area has very low susceptibility, mostly in the lower and eastern parts.

Salinity (both in soil and groundwater) is a major factor in physical susceptibility to drought. The area with high susceptibility primarily has saline soil and poor-quality groundwater with low expected yield. This is mainly due to the rock salt strata under the basin (Wongsomsak, 1986). Generally, soils in Buri Ram, Surin, and Si Sa Ket provinces have better soil water-holding capacity, which diminishes the susceptibility;

however, in some plots, susceptibility rises to moderate and high because of saline soils and saline groundwater with low expected yield. The eastern and lower parts are not affected by salt, and has a high yield and good quality groundwater, although the soil water-holding capacities in some plots are low. This leads to low and very low susceptibility in these areas.

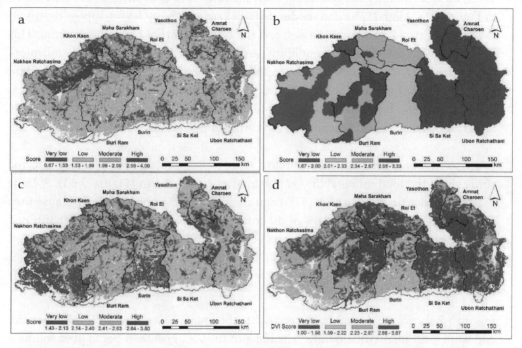

Figure 5-6 Physical (a), socioeconomic (b) and total (c) susceptibility, and total vulnerability (d).

In terms of socioeconomic susceptibility, the eastern part of the basin and the provinces in the central part, except Surin, are more susceptible to drought than the others (Figure 5-6b). High (45%) and moderate (14%) levels of susceptibility are mostly found in the eastern region and some districts in the central region, whereas low (25%) and very low (16%) susceptibilities are primarily in the western region and Surin. Nakhon Ratchasima shows very low and low susceptibility.

The socioeconomic susceptibility would be higher if the poverty ratio of the farmers was used in the socioeconomic factors. The poverty ratio used in this study is based on the total population of the province, not only farmers. In Thailand, farmers, particularly the small farm holders, generally earn lower incomes than people in other occupations. In the northeast, the average wage in the agriculture sector is usually the lowest ($165/month in 2017) (National Statistical Office Thailand, 2018). In addition, for small farm holders, rice production is primarily considered only for self-consumption because the income from rice cultivation is insufficient to cover household expenses or to provide working capital for the next growing season (Bank of Thailand, 2015; Haefele et al., 2006).

The total susceptibility integrated both physical and socioeconomic factors is depicted in Figure 5-6c. High (12%) and moderate (25%) susceptibility levels appear mostly in the lower parts – Buri Ram, Si Sa Ket, and the western part of Ubon Ratchathani. This is because of the high susceptibility associated with socioeconomic indicators dominates in those areas. However, in the rest of the basin, particularly in Nakhon Ratchasima, Surin, and Khon Kaen, the susceptibility due to physical indicators is lessened to lower levels (36% low susceptibility and 27% very low) because of the low levels associated with socioeconomic indicators in those areas.

When combining the total susceptibility with coping and adaptive capacity in the study area, the total vulnerability is obtained as presented in (Figure 5-6d). A large majority of rice fields have high (57%) and moderate (38%) vulnerability because they are rain-fed agriculture. Only 8% of the total rice cultivation is irrigated (distributed over the basin) resulting in low (3%) and very low (2%) vulnerability in those irrigated areas.

5.3.4 Risk

Figure 5-7 shows the drought risks rice cultivation faces based on the monthly hazard, exposure, and total vulnerability. The risks in July, August, and September follow a similar pattern as the hazard maps, increasing from east to west. In those months, the risk is mostly low and very low; only small areas are at moderate risk. September presents the lowest risk (72% of areas have very low risk) because of the high precipitation. In contrast, the risks in October and November become more severe and show the same pattern, decreasing from north to south. Approximately 65% of the area shows moderate risk, with the highest levels at the central part of the basin, and the lower and eastern parts with the lowest levels.

The risk map of total precipitation over the entire growing season does not identify areas with critical conditions as the monthly time step does. This map demonstrates that about 94% of rice fields have low and very low risks, while the remaining 6%, mostly in Nakhon Ratchasima and Buri Ram, have moderate risk. This overall perspective is significantly different from the month-to-month perspective, in which one can see that some specific dry periods can coincide with vital growth stages, leading to significantly reduced yields. If we only consider the total precipitation, it can distort our perception and may result in improper decision-making in certain places. Moreover, two important stages, flowering, and grain filling, both of which influence the yield, occur in October. If the initial planting is delayed, delaying the onset of the growing season, these vital stages could be postponed to November, putting them at risk of serious water shortage.

The average education level of people in all provinces, except in Khon Kaen, is below the current Thai national compulsory education of 9 years (Office of the Education Council, 2014), thus improving education level has the potential to reduce drought vulnerability in the region. A UNDP report (Pelling et al., 2004) indicates that better-educated people are better able to cooperate and collaborate with experts in designing ways of dealing with and mitigating disaster risk and respond better to warnings and other public announcements. Moreover, better-educated people can more easily access information about hazard preparedness, reduction, and adaptation, which are increasingly available through new technologies. Well-informed people can help to reduce disaster risks (O.D. Cardona et al., 2012).

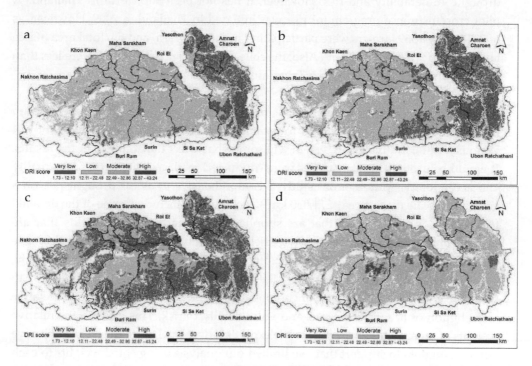

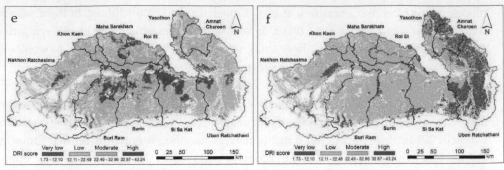

Figure 5-7 Drought risk map of month Jul (a), Aug (b), Sep (c), Oct (d), Nov (e) and total precipitation over the entire rice-growing season (f).

Crop insurance, a mitigation measure that compensates farmers for crop losses due to natural disasters (Wilhelmi & Wilhite, 2002), is another crucial factor that can reduce drought vulnerability and risk. However, it has not been successful in Thailand. A pilot crop insurance program for rice was initiated in Thailand in 2009. However, as of 2014, only 1369 farmers were participating in the program, and the total area of rice land insured was only 45.3 km². Also, the collected insurance premiums were less than the indemnity payments (Sinha & Tripathi, 2016).

Since this study considers numerous factors, data quality and availability are its main limitation. The socioeconomic data, for example, are provided in the averaged form at various administrative levels, which will somewhat affect the accuracy of the results. It would be useful if similar studies were conducted in places with better data quality and accessibility. Although our study is comprehensive to assess drought risk on a river basin scale (basin area of 71,060 km²), we have not included some of the detailed factors, e.g., irrigated areas that are supplied by groundwater, rice fields that are insured under crop insurance, and rice fields where drought-tolerant varieties have been planted. Future studies should explore the significance of these factors on the basin-wide assessment. Furthermore, the relative contribution of each indicator to exposure and vulnerability or their components is space-specific, and so is the relative contribution of hazard, exposure, and vulnerability to risk. This leads to the relative weights of these factors subjective to the expert's judgment. Due to the lack of objective criteria and data to support that, we limited our analysis using equal weights to each factor. Further analysis of relative weights should carry out the sensitivity of risk factors and may prepare multiple risk maps with various weight scenarios.

5.4 Conclusions

This paper aimed to identify the spatial variations in drought hazard, exposure, vulnerability, and risk of rice cultivation at the Mun River Basin, Thailand using a comprehensive approach, which includes three key components of risk, i.e., hazard, exposure, and vulnerability. The drought hazard was assessed by comparing water deficit to the water requirements of the rice crop as an indicator. The frequency of drought occurrences was directly used to derive DHI, which is more sensitive to changes in drought severity. The analysis was carried out at monthly time steps, highlighting dynamics of spatial variations in hazard and risk. The exposed population and rice were included in the exposure domain, while the vulnerability domain consists of physical, socioeconomic, and coping and adaptive capacity units.

The findings from this study highlight: (1) the importance of assessing agricultural drought hazard and risk on a monthly basis, because total precipitation over the rice-growing season does not reflect the actual conditions the rice has experienced, which can cause considerable reductions in yield; and (2) the importance of including all relevant components, i.e., hazard (severity and frequency), exposure (exposed people and elements), and vulnerability (physical, socioeconomic, and coping and adaptive capacity), into the analysis when conducting drought risk assessment in order to understand risk in all dimension. We showed that the hazard and risk maps of total precipitation indicate lower levels of drought hazard and risk, with different patterns than those of specific months. Consequently, we recommend adopting the hazard and risk maps for October and November as the basis for developing solutions to mitigate against drought impacts yield in the basin.

The contribution of the study is twofold. (1) It helps to understand the water conditions, exposure, vulnerability, and drought risk facing rice cultivation in the Mun River Basin in Thailand. This information is essential for all relevant stakeholders, the government, water-related authorities, regulators, policymakers, as well as farmers, to improve the yields by means of improved water resource management and coping and adaptive capacity, as well as social and economic development. (2) It offers a comprehensive drought risk assessment scheme and the method of hazard analysis specific to rice, which can be applied to other basins or areas with single or multi-crops.

References

Allen, R. G., Pereira, L. S., Raes, D., & Smith, M. (1998). FAO Irrigation and Drainage Paper No. 56, Crop Evapotranspiration (Guidelines for Computing Crop Water Requirements). FAO. *Water Resources, Development and Management Service, Rome, Italy.*

Antwi-Agyei, P., Fraser, E. D., Dougill, A. J., Stringer, L. C., & Simelton, E. (2012). Mapping the vulnerability of crop production to drought in Ghana using rainfall, yield and socioeconomic data. *Applied Geography, 32*(2), 324-334.

Bank of Thailand. (2015). *The effect of government subsidy to farmers (1000 baht per rai) on the economic growth in Northeast Thailand (in Thai).* Retrieved from Bangkok, Thailand: www.bot.or.th

Beighley, D. H. (2010). Growth and production of rice. *Soils, plant growth and crop production, Encyclopedia of Life Support Systems (EOLSS), 2,* 349.

Belal, A.-A., El-Ramady, H. R., Mohamed, E. S., & Saleh, A. M. (2014). Drought risk assessment using remote sensing and GIS techniques. *Arabian Journal of Geosciences, 7*(1), 35-53.

Blauhut, V., Gudmundsson, L., & Stahl, K. (2015). Towards pan-European drought risk maps: quantifying the link between drought indices and reported drought impacts. *Environmental research letters, 10*(1), 014008.

Bouman, B. A. M., Lampayan, R. M., & Tuong, T. P. (2007). *Water management in irrigated rice: coping with water scarcity.* Los Banos, Philippines: International Rice Research Institute.

Brouwer, C., Goffeau, A., & Heibloem, M. (1985). Irrigation Water Management: Training Manual No. 1-Introduction to Irrigation. *Food and Agriculture Organization of the United Nations, Rome, Italy.*

Brouwer, C., Prins, K., & Heibloem, M. (1989). Irrigation Water Management: Irrigation Scheduling. In *Irrigation water management: Training manual no. 4.* Rome, Italy: Food and Agriculture Organization of the United Nations (FAO).

Bureau of Rice Research and Development (BRRD). (n.d., 14 July 2016). Rice Knowledge Bank (in Thai). *Ministry of Agriculture and Cooperatives, Thailand.* Retrieved from http://www.brrd.in.th/rkb/

Cardona, O. D. (2011). Disaster risk and vulnerability: Concepts and measurement of human and environmental insecurity. In H. G. Brauch, Ú. O. Spring, C. Mesjasz, J. Grin, P. Kameri-Mbote, B. Chourou, & J. B. Pál Dunay (Eds.), *Coping with Global Environmental Change, Disasters and Security* (pp. 107-121): Springer.

Cardona, O. D., van Aalst, M. K., Birkmann, J., Fordham, M., McGregor, G., Perez, R., . . . Sinh, B. T. (2012). Determinants of risk: exposure and vulnerability. In C. B. Field, V. Barros, T. F. Stocker, D. Qin, D. J. Dokken, K. L. Ebi, M. D. Mastrandrea, K. J. Mach, G. K. Plattner, S. K. Allen, M. Tignor, & P. M. Midgley (Eds.),

Managing the Risks of Extreme Events and Disasters to Advance Climate Change Adaptation. A Special Report of Working Groups I and II of the Intergovernmental Panel on Climate Change (IPCC) (pp. 65-108). Cambridge, UK, and New York, NY, USA: Cambridge University Press.

Carrão, H., Naumann, G., & Barbosa, P. (2016). Mapping global patterns of drought risk: An empirical framework based on sub-national estimates of hazard, exposure and vulnerability. *Global Environmental Change, 39*, 108-124. doi:https://doi.org/10.1016/j.gloenvcha.2016.04.012

Chen, W., Cutter, S. L., Emrich, C. T., & Shi, P. (2013). Measuring social vulnerability to natural hazards in the Yangtze River Delta region, China. *International Journal of Disaster Risk Science, 4*(4), 169-181.

Chung, N. T., Jintrawet, A., & Promburom, P. (2015). Impacts of seasonal climate variability on rice production in the central highlands of Vietnam. *Agriculture and Agricultural Science Procedia, 5*, 83-88.

Dah-gbeto, A. P., & Villamor, G. B. (2016). Gender-specific responses to climate variability in a semi-arid ecosystem in northern Benin. *Ambio, 45*(3), 297-308.

Daneshvar, M. R. M., Bagherzadeh, A., & Khosravi, M. (2013). Assessment of drought hazard impact on wheat cultivation using standardized precipitation index in Iran. *Arabian Journal of Geosciences, 6*(11), 4463-4473.

De Smith, M. J., Goodchild, M. F., & Longley, P. (2009). *Geospatial analysis: a comprehensive guide to principles, techniques and software tools* (3rd ed.). Leicester LE1 7FW, UK: Matador.

Department of Water Resources of Thailand. (2016). *Summary of the results of drought prevention and mitigation year 2015-2016. Final Report (in Thai)*. Bangkok, Thailand.

Djoudi, H., & Brockhaus, M. (2011). Is adaptation to climate change gender neutral? Lessons from communities dependent on livestock and forests in northern Mali. *International Forestry Review, 13*(2), 123-135.

Eriyagama, N., Smakhtin, V., Chandrapala, L., & Fernando, K. (2010). *Impacts of climate change on water resources and agriculture in Sri Lanka: a review and preliminary vulnerability mapping* (Vol. 135): IWMI.

FAO. (2009, January 2009). ETo Calculator. Retrieved from http://www.fao.org/land-water/databases-and-software/eto-calculator/en/

FAO. (2017). *Drought characterization and management in Central Asia Region and Turkey*. Retrieved from http://www.fao.org/documents/card/en/c/d2da11f3-4d0c-4f30-ab8d-fe6a0cd348ab/

FAO. (2018). Drought and Agriculture. Retrieved from http://www.fao.org/land-water/water/drought/droughtandag/en/

Fukai, S., Basnayake, J., & Cooper, M. (2000). Modelling water availability, crop growth, and yield of rainfed lowland rice genotypes in northeast Thailand.

Characterising and understanding rainfed environments. Los Baños, Philippines, IRRI, 111-130.

Haefele, S., Naklang, K., Harnpichitvitaya, D., Jearakongman, S., Skulkhu, E., Romyen, P., . . . Khunthasuvon, S. (2006). Factors affecting rice yield and fertilizer response in rainfed lowlands of northeast Thailand. *Field crops research, 98*(1), 39-51.

Hagenlocher, M., Renaud, F. G., Haas, S., & Sebesvari, Z. (2018). Vulnerability and risk of deltaic social-ecological systems exposed to multiple hazards. *Science of the Total Environment, 631,* 71-80.

Hayes, M. J., Wilhelmi, O. V., & Knutson, C. L. (2004). Reducing drought risk: bridging theory and practice. *Natural Hazards Review, 5*(2), 106-113.

He, B., Lü, A., Wu, J., Zhao, L., & Liu, M. (2011). Drought hazard assessment and spatial characteristics analysis in China. *Journal of Geographical Sciences, 21*(2), 235-249.

He, B., Wu, J., Lü, A., Cui, X., Zhou, L., Liu, M., & Zhao, L. (2013). Quantitative assessment and spatial characteristic analysis of agricultural drought risk in China. *Natural Hazards, 66*(2), 155-166.

IPCC. (2012). *Managing the Risks of Extreme Events and Disasters to Advance Climate Change Adaptation. A Special Report of Working Groups I and II of the Intergovernmental Panel on Climate Change.* Cambridge, UK, and New York, NY, USA: Cambridge University Press.

IPCC. (2014). Summary for policymakers. In C. B. Field, V.R. Barros, D.J. Dokken, K.J. Mach, M.D. Mastrandrea, T.E. Bilir, M. Chatterjee, K.L. Ebi, Y.O. Estrada, R.C. Genova, B. Girma, E.S. Kissel, A.N. Levy, S. MacCracken, P.R. Mastrandrea, & L. L. White (Eds.), *Climate change 2014: Impacts, Adaptation, and Vulnerability. Part A: Global and Sectoral Aspects. Contribution of Working Group II to the Fifth Assessment Report of the Intergovernmental Panel on Climate Change* (pp. 1-32). Cambridge, United Kingdom and New York, NY, USA: Cambridge University Press.

Jearakongman, S., Rajatasereekul, S., Naklang, K., Romyen, P., Fukai, S., Skulkhu, E., . . . Nathabutr, K. (1995). Growth and grain yield of contrasting rice cultivars grown under different conditions of water availability. *Field crops research, 44*(2), 139-150.

Jeong, D. I., Sushama, L., & Naveed Khaliq, M. (2014). The role of temperature in drought projections over North America. *Climatic Change, 127*(2), 289-303. doi:10.1007/s10584-014-1248-3

Keyantash, J. A., & Dracup, J. A. (2004). An aggregate drought index: Assessing drought severity based on fluctuations in the hydrologic cycle and surface water storage. *Water Resources Research, 40*(9), 1-13.

Kim, H., Park, J., Yoo, J., & Kim, T.-W. (2015). Assessment of drought hazard, vulnerability, and risk: a case study for administrative districts in South Korea. *Journal of Hydro-environment Research, 9*(1), 28-35.

Knutson, C., Hayes, M., & Phillips, T. (1998). How to reduce drought risk. A guide prepared by the preparedness and mitigation working group of the Western Drought Coordination Council. National Drought Mitigation Center, Lincoln, Nebraska. *drought. unl. edu/plan/handbook/risk. pdf.*

Koks, E. E., Jongman, B., Husby, T. G., & Botzen, W. J. (2015). Combining hazard, exposure and social vulnerability to provide lessons for flood risk management. *environmental science & policy, 47*, 42-52.

Kron, W. (2002). *Keynote lecture: Flood risk= hazard× exposure× vulnerability.* Paper presented at the Flood defence, Beijing.

Kron, W. (2005). Flood risk= hazard• values• vulnerability. *Water international, 30*(1), 58-68.

Lei, Y., & Luo, L. (2011). Drought risk assessment of China's mid-season paddy. *International Journal of Disaster Risk Science, 2*(2), 32-40.

Li, T., Angeles, O., Radanielson, A., Marcaida, M., & Manalo, E. (2015). Drought stress impacts of climate change on rainfed rice in South Asia. *Climatic Change, 133*(4), 709-720.

Liu, X., Wang, Y., Peng, J., Braimoh, A. K., & Yin, H. (2013). Assessing vulnerability to drought based on exposure, sensitivity and adaptive capacity: a case study in middle Inner Mongolia of China. *Chinese Geographical Science, 23*(1), 13-25.

Lloyd-Hughes, B., & Saunders, M. A. (2002). A drought climatology for Europe. *International Journal of Climatology, 22*(13), 1571-1592.

Maracchi, G. (2000). Agricultural drought−a practical approach to definition, assessment and mitigation strategies. In J. V. Vogt & F. Somma (Eds.), *Drought and drought mitigation in Europe* (Vol. 14, pp. 63-75): Springer.

Mongkolsawat, C., Thirangoon, P., Suwanweramtorn, R., Karladee, N., Paiboonsank, S., & Champathet, P. (2001). An evaluation of drought risk area in Northeast Thailand using remotely sensed data and GIS. *Asian Journal of Geoinformatics, 1*(4), 33-44.

Murthy, C., Laxman, B., & Sai, M. S. (2015). Geospatial analysis of agricultural drought vulnerability using a composite index based on exposure, sensitivity and adaptive capacity. *International journal of disaster risk reduction, 12*, 163-171.

National Statistical Office Thailand. (2018). *Average Wage by Economic Sector in the Northeast Thailand (in Thai).* Retrieved from: http://www2.bot.or.th/statistics/ReportPage.aspx?reportID=740&language= th

Naumann, G., Barbosa, P., Garrote, L., Iglesias, A., & Vogt, J. (2014). Exploring drought vulnerability in Africa: an indicator based analysis to be used in early warning systems. *Hydrology and Earth System Sciences, 18*(5), 1591-1604.

Nawata, E., Nagata, Y., Sasaki, A., Iwama, K., & Sakuratani, T. (2005). Mapping of climatic data in Northeast Thailand: Rainfall. *Tropics, 14*(2), 191-201.

Office of the Education Council. (2014). *Map of Provincial Development of Thailand: Education (in Thai)* (Vol. 1). Bangkok, Thailand: Office of the Education Council, Ministry of Education.

Pandey, R. P., Pandey, A., Galkate, R. V., Byun, H.-R., & Mal, B. C. (2010). Integrating hydro-meteorological and physiographic factors for assessment of vulnerability to drought. *Water Resources Management, 24*(15), 4199-4217.

Pandey, S., Pandey, A., Nathawat, M., Kumar, M., & Mahanti, N. (2012). Drought hazard assessment using geoinformatics over parts of Chotanagpur plateau region, Jharkhand, India. *Natural Hazards, 63*(2), 279-303.

Peduzzi, P., Dao, H., Herold, C., & Mouton, F. (2009). Assessing global exposure and vulnerability towards natural hazards: the Disaster Risk Index. *Natural Hazards and Earth System Sciences, 9*(4), 1149-1159.

Pelling, M., Maskrey, A., Ruiz, P., Hall, P., Peduzzi, P., Dao, Q.-H., . . . Kluser, S. (2004). *Reducing disaster risk: a challenge for development*. Retrieved from New York: United Nations:

Pereira, L. S., Cordery, I., & Iacovides, I. (2002). *Coping with water scarcity*: Springer.

Prabnakorn, S., Maskey, S., Suryadi, F., & de Fraiture, C. (2016, 6-8 Nov 2016). *Climate and Drought Trends and Their Relationships with Rice Production in the Mun River Basin, Thailand*. Paper presented at the International Commission on Irrigation and Drainage (ICID) Conference Proceedings, Chiang Mai, Thailand.

Prabnakorn, S., Maskey, S., Suryadi, F., & de Fraiture, C. (2018). Rice yield in response to climate trends and drought index in the Mun River Basin, Thailand. *Science of the Total Environment, 621*, 108-119. doi:10.1016/j.scitotenv.2017.11.136

Rahman, M. A., Kang, S., Nagabhatla, N., & Macnee, R. (2017). Impacts of temperature and rainfall variation on rice productivity in major ecosystems of Bangladesh. *Agriculture & Food Security, 6*(1), 10.

Safavi, H. R., Esfahani, M. K., & Zamani, A. R. (2014). Integrated index for assessment of vulnerability to drought, case study: Zayandehrood River Basin, Iran. *Water Resources Management, 28*(6), 1671-1688.

Sarker, M. A. R., Alam, K., & Gow, J. (2012). Exploring the relationship between climate change and rice yield in Bangladesh: An analysis of time series data. *Agricultural Systems, 112*, 11-16.

Schindler, U., Steidl, J., Müller, L., Eulenstein, F., & Thiere, J. (2007). Drought risk to agricultural land in Northeast and Central Germany. *Journal of Plant Nutrition and Soil Science, 170*(3), 357-362.

Sebesvari, Z., Renaud, F. G., Haas, S., Tessler, Z., Hagenlocher, M., Kloos, J., . . . Kuenzer, C. (2016). A review of vulnerability indicators for deltaic social–ecological systems. *Sustainability Science, 11*(4), 575-590.

Shahid, S., & Behrawan, H. (2008). Drought risk assessment in the western part of Bangladesh. *Natural Hazards, 46*(3), 391-413.

Sinha, S., & Tripathi, N. K. (2016). Assessing the Challenges in Successful Implementation and Adoption of Crop Insurance in Thailand. *Sustainability, 8*(12), 1306.

Spinoni, J., Naumann, G., Carrao, H., Barbosa, P., & Vogt, J. (2014). World drought frequency, duration, and severity for 1951–2010. *International Journal of Climatology, 34*(8), 2792-2804.

Swain, M., & Swain, M. (2011). Vulnerability to agricultural drought in Western Orissa: A case study of representative blocks. *Agricultural Economics Research Review, 24*(347-2016-16890), 47-56.

Tsakiris, G., Pangalou, D., & Vangelis, H. (2007). Regional Drought Assessment Based on the Reconnaissance Drought Index (RDI). *Water Resources Management, 21*(5), 821-833. doi:10.1007/s11269-006-9105-4

Wang, Q., Wu, J., Lei, T., He, B., Wu, Z., Liu, M., . . . Zhou, H. (2014). Temporal-spatial characteristics of severe drought events and their impact on agriculture on a global scale. *Quaternary International, 349*, 10-21.

Wilhelmi, O. V., & Wilhite, D. A. (2002). Assessing vulnerability to agricultural drought: a Nebraska case study. *Natural Hazards, 25*(1), 37-58.

Wilhite, D. A. (2000). Drought as a natural hazard: Concepts and definitions. In D. A. Wilhite (Ed.), *Drought, a global assessment* (Vol. 1, pp. 1-18). New York: Routledge.

Wongsomsak, S. (1986). Salinization in Northeast Thailand (< Special Issue> Problem Soils in Southeast Asia).

Wopereis, M., Defoer, T., Idinoba, P., Diack, S., & Dugué, M. (2008). Participatory learning and action research (PLAR) for integrated rice management (IRM) in inland valleys of sub-Saharan Africa: technical manual. *WARDA Training Series. Africa Rice Center, Cotonou, Benin, 128*, 26-32.

World Meteorological Organization (WMO) and Global Water Partnership (GWP). (2016). Handbook of Drought Indicators and Indices In M. Svoboda & B. A. Fuchs (Eds.), *Integrated Drought Management Programme (IDMP), Integrated Drought Management Tools and Guidelines Series 2*. Geneva, Switzerland.

Wu, H., Hubbard, K. G., & Wilhite, D. A. (2004). An agricultural drought risk-assessment model for corn and soybeans. *International Journal of Climatology, 24*(6), 723-741.

Wu, H., & Wilhite, D. A. (2004). An operational agricultural drought risk assessment model for Nebraska, USA. *Natural Hazards, 33*(1), 1-21.

Wu, J., He, B., Lü, A., Zhou, L., Liu, M., & Zhao, L. (2011). Quantitative assessment and spatial characteristics analysis of agricultural drought vulnerability in China. *Natural Hazards, 56*(3), 785-801.

Yaduvanshi, A., Srivastava, P. K., & Pandey, A. C. (2015). Integrating TRMM and MODIS satellite with socio-economic vulnerability for monitoring drought risk over a tropical region of India. *Physics and Chemistry of the Earth, Parts A/B/C, 83-84*, 14-27. doi:https://doi.org/10.1016/j.pce.2015.01.006

Yevjevich, V. (1967). *An objective approach to definitions and investigations of continental hydrologic droughts*: Colorado State University, Fort Collins, Colorado.

Yoshida, S. (1981). *Fundamentals of rice crop science*. Los Banos, Philippines: Internation Rice Research Institute.

Yuan, X.-C., Wang, Q., Wang, K., Wang, B., Jin, J.-L., & Wei, Y.-M. (2015). China's regional vulnerability to drought and its mitigation strategies under climate change: data envelopment analysis and analytic hierarchy process integrated approach. *Mitigation and Adaptation Strategies for Global Change, 20*(3), 341-359.

Zhang, Q., Sun, P., Li, J., Xiao, M., & Singh, V. P. (2015). Assessment of drought vulnerability of the Tarim River basin, Xinjiang, China. *Theoretical and applied climatology, 121*(1-2), 337-347.

Zimmerman, D., Pavlik, C., Ruggles, A., & Armstrong, M. P. (1999). An experimental comparison of ordinary and universal kriging and inverse distance weighting. *Mathematical Geology, 31*(4), 375-390.

Appendix

Table A-5.1 provides a list of factors for vulnerability assessment to drought and agricultural drought summarizing from some scientific papers through a literature review. We selected the peer-reviewed papers from Google Scholar and the search criteria included relevant terms such as "drought", "vulnerability", "risk" because vulnerability is part of risk ("allintitle: drought vulnerability risk", and "allintitle: drought risk"), "agriculture" to specify areas of interest ("allintitle: drought vulnerability agriculture OR agricultural OR crop", and "drought risk agriculture OR agricultural OR crop"), and "assessment" terms to identify research related to vulnerability assessment ("allintitle: drought vulnerability assessment OR accessing OR analysis"). The papers from year 2000 are high priority but ones that consider risk or vulnerability based on a hazard-centric approach or their final results not in form of spatial distribution are excluded because they are out of line with our study. The papers that consider risk or vulnerability based on a hazard-centric approach or their final results not in form of spatial distribution are excluded because they are out of line with our study. The factors are classified into three groups, i.e., biophysical, socio-economic and coping capacity that usually shown in these reviewed papers.

Table A- 5.1   Weights for different levels of water deficit

Indicator	Factor	Wilhelmi and Wilhite (2002)	Shahid and Behrawan (2008)	Eriyagama, Smakhtin, Chandrapala, and Fernando (2010)	J. Wu et al. (2011) ; He et al. (2013)	Swain and Swain (2011)	Antwi-Agyei, Fraser, Dougill, Stringer, and Simelton (2012)	Murthy, Laxman, and Sai (2015)	Zhang, Sun, Li, Xiao, and Singh (2015)	R. P. Pandey, Pandey, Galkate, Byun, and Mal	Liu et al. (2013)	Naumann, Barbosa, Garrote, Iglesias, and Vogt (2014)	Safavi et al. (2014)	Kim et al. (2015)	Yaduvanshi, Srivastava, and Pandey (2015)	Yuan et al. (2015)	Carrão et al. (2016)
		Agricultural drought										Drought					
Physical	Land use**	+		+		+		+		+		+	+			+	
	Soil water-holding capacity (mm)	-	-		-	-		-		-			-				
	Crop production (t/ha, kg/ha) Yield variability (%)			+		+	+		+					+			
	Total water use (% of renewable)											-					-
	Soil salinity (%)																
	Land slope (%)					+				+				+			
	Groundwater availability/quantity (m³, litre)					-				-			-				

Indicator	Factor	Wilhelmi and Wilhite (2002)	Shahid and Behrawan (2008)	Eriyagama, Smakhtin, Chandrapala, and Fernando (2010)	J. Wu et al. (2011); He et al. (2013)	Swain and Swain (2011)	Antwi-Agyei, Fraser, Dougill, Stringer, and Simelton (2012)	Murthy, Laxman, and Sai (2015)	Zhang, Sun, Li, Xiao, and Singh (2015)	R. P. Pandey, Pandey, Galkate, Byun, and Mal	Liu et al. (2013)	Naumann, Barbosa, Garrote, Iglesias, and Vogt (2014)	Safavi et al. (2014)	Kim et al. (2015)	Yaduvanshi, Srivastava, and Pandey (2015)	Yuan et al. (2015)	Carrão et al. (2016)
		Agricultural drought								Drought							
	Surface water availability/quantity (m³, litre)									-			-				
	Environmental needs (m³, litre)												+				
	Percent of crop area with small holding (%)								+								
	Water utilization by sector (%) (agriculture, municipality, industry, etc.)									+		+		+		+	
	Cultivated land area per capita (ha, km², etc.)									-							
Socio-economic	Population density (people/ha, people/km², etc.)	+				+			+	+		+		+	+	+	
	Female to male ratio (%)	+							+						+		
	Poverty level (%)	+	+		+	+						+					+
	Poverty gap ratio (%)			+													
	Gross domestic product (GDP) per capita (US$, €)											-	-			-	-
	Saving per capita (US$, €)												-				
	People depending on agriculture (%) Agricultural employment (%)	+	+						+					+	+		
	Landless labourer to total main workers (%)						+										
	Agriculture (% of GDP)			+					+	+		+					+
	Education/Literacy rate (%, year of schooling)			-		-	-		-			-			-		-
	Life expectancy at birth (years)											-					-
	People living in rural area (%, people per grid cell)			+		+						+				+	+
	Population ages 15-64 (%)																-
	Dependency ratio (%) Children < 5 years (%) Population in retirement (%)								+								
	Population with access to improved water (%)											-					-
	Energy use (kg oil equivalent per person, million Btu per person)			-								-					-

Indicator	Factor	Wilhelmi and Wilhite (2002)	Shahid and Behrawan (2008)	Eriyagama, Smakhtin, Chandrapala, and Fernando (2010)	J. Wu et al. (2011) ; He et al. (2013)	Swain and Swain (2011)	Antwi-Agyei, Fraser, Dougill, Stringer, and Simelton (2012)	Murthy, Laxman, and Sai (2015)	Zhang, Sun, Li, Xiao, and Singh (2015)	R. P. Pandey, Pandey, Galkate, Byun, and Mal	Liu et al. (2013)	Naumann, Barbosa, Garrote, Iglesias, and Vogt (2014)	Safavi et al. (2014)	Kim et al. (2015)	Yaduvanshi, Srivastava, and Pandey (2015)	Yuan et al. (2015)	Carrão et al. (2016)
						Agricultural drought							Drought				
	Refugee population or territory of asylum (% of total population)											+					+
Coping capacity	Irrigation (% of total agriculture, km² per grid cell, have/not have)	-	-	-	-	-		-	-			-		-		-	-
	Area/Farmer insured under crop insurance (%)						-										
	Fertilizer consumption (kg/ha)											-					
	People under special support of the government (%) / Government effectiveness**						-					-					-
	Infrastructure for prevention & preparedness (US$/year/capita, %storage/total water use)											-				-	-
	Road density (km of road per 100 sq.km of land area)	-															-
	Communication index (e.g. the number of landlines, internet users)	-									-						
	Institutional capacity**											-					
	The number of technologists per 1000 persons											-					

*Indicates if high indicator scores increase (+) or decrease (-) vulnerability; **no unit - the author gives a certain range (or score) to the indicator.

Table A- 5.2   Correlation matrix of exposed rice

Factor		LU	RP
Land use type: LU	Kendall's tau	1.000	-0.162
	Sig. (2-tailed)		0.227
Rice production: RP	Kendall's tau	-0.162	1.000
	Sig. (2-tailed)	0.227	

Conclusion: No issue of multicollinearity detected.

Table A- 5.3   Correlation matrix of physical susceptibility

Factor		SM	SL	GW
Soil water-holding capacity:	Kendall's tau	1.000	-0.061	-0.050
	Sig. (2-tailed)		0.628	0.676
Saline soil level: SL	Kendall's tau	-0.061	1.000	0.274*
	Sig. (2-tailed)	0.628		0.021
Groundwater quantity: GW	Kendall's tau	-0.050	0.274	1.000
	Sig. (2-tailed)	0.676	0.021	

Conclusion: No issue of multicollinearity detected.

Table A- 5.4   Correlation matrix of social susceptibility.

Factor		PV	PA	ED
Poverty: PV	Kendall's tau	1.000	-0.196	0.192
	Sig. (2-tailed)		0.100	0.137
People depending on	Kendall's tau	-0.196	1.000	0.053
	Sig. (2-tailed)	0.100		0.681
Education level – average year	Kendall's tau	0.192	0.053	1.000
	Sig. (2-tailed)	0.137	0.681	

Conclusion: No issue of multicollinearity detected.

The red and blue colors represent negative and positive values respectively.

Flood and drought mitigation measures and strategies

This chapter is based on:
Prabnakorn S, Ruangpan L, Suryadi F, de Fraiture C. (2019). Solving floods and droughts for rice cultivation at the Mun River Basin in Thailand. (*Submitted*).

Abstract

Thailand regularly experiences flooding in the wet season and drought in the dry season, and damage from both hazards are expected to increase in the future. Agriculture, particularly rice, is one of the main sectors affected. Thus, a massive amount of money has been spent on water resources development projects aiming at mitigating flood problems and improving water availability in the dry season, particularly for agricultural purposes. However, the results are not as intended, as both hazards still regularly occur. This study aims to solve flood and drought damage to rice cultivation in the Mun River Basin in Thailand by examining all dimensions regarding flood and drought problems to rice cultivation at the basin scale, including their adverse impacts, and the coping capacity of existing measures. The results demonstrate that the total storage capacity of the in-situ and ongoing projects is sufficient to tackle both hazards, but their performance is inefficient and ineffective. We thus offer several suggestions to improve the performance of existing measures. To complement those and make the storage systems more flexibility, we propose other measures: farm pond with the appropriate size occupies about 10% of the farm area, subsurface floodwater harvesting system, and oxbow lake reconnection. The suggested measures are practicable, economical, and less environmental impacts; they, therefore, have the potential to be implemented. If they are executed with appropriate pre- and post-project reviews, the flood and drought problems in the basin could be solved or reduced sustainably. Moreover, the equation for estimating the size of the farm pond that reflects the actual demand for wet-season rice can be applied to design ponds for other crops or other specified periods.

Keywords: Agriculture, Farm pond, the MAR, Water demand, Water supply

6.1 Introduction

Thailand has periodically suffered from flooding during the wet season and from drought during the dry season Pavelic et al. (2012); (Shannon, 2005). Flooding is the most frequent natural disaster affected Thailand, 66 floods over the period of 1984 – 2014 causing the total damage of US 44,885 million approximately. Riverine flood is the most common one, which is usually induced by monsoonal and torrential rains and sometimes by tropical storms (Centre for Research on the Epidemiology of Disaster (CRED), 2015). Whereas at the same time, the country has regularly been experienced droughts owing to high seasonal variability of rainfall, which finally leads to widespread crop failure. The areas at greatest risk to drought are in the northeast region, i.e., the Chi and Mun River Basins. The cumulative damage from droughts between 1989 and 2014 was estimated to US 1,143 million (Department of Water Resources of Thailand, 2016).

Adverse impacts from flood and drought in Thailand are on the rise. The temperature tends to increase by 0.2 – 0.3 °C per decade in both warm and cold seasons (Johnston et al., 2010). Specifically, in the Mun River Basin in the northeast, the minimum and maximum temperatures show the upward trends with the rates up to 0.1 – 0.65 °C and 0.8 °C per decade, respectively (Prabnakorn, Maskey, Suryadi, & de Fraiture, 2018). Further, a consistent rise in the number of the consecutive dry day is observed, especially the eastern part of the basin (Artlert & Chaleeraktrakoon, 2013; Manomaiphiboon, Octaviani, Torsri, & Towprayoon, 2013). The dry season will, therefore, be dryer and longer (Snidvongs, Choowaew, & Chinvanno, 2003). This will increase evapotranspiration and water demand for agriculture, making crops more vulnerable to yield reduction. For flood, although there is a controversy over a projection of rainfall variations, and the trends are mixed, there is a common conclusion that heavy precipitation events will become more intense and more frequent (Bates, Kundzewicz, Wu, & Palutikof, 2008; Johnston et al., 2010; Manomaiphiboon et al., 2013; Snidvongs et al., 2003). Along with a high degree of urbanization, demographic shifts, and land-use changes account for the increase in flood damage (Johnston et al., 2010).

As a major exporter of rice, agriculture, particularly rice, is the primary sector to be affected. Rice is the most important staple crop of Thailand, occupying almost half of the agricultural land of the country, and approximately 60% of that locates in the north-eastern region. Most of the rice fields in the northeast are traditional farming, under rain-fed conditions, and highly susceptible to climate variations. With the least proportion of rice fields under irrigation support resulted in the lowest yield of the country (2.3 ton/ha) comparing to 3.6, 3.9, 3 ton/ha in the north, the central, and the south, respectively, and also below the country's average yield at 2.8 ton/ha (Office of Agricultural Economics, 2018; Prabnakorn, Maskey, Suryadi, & de Fraiture, 2016).

That shows the need for water development projects to safeguard rice agriculture from floods and droughts. The government, therefore, has spent billions of dollars in increasing water storage capacity aiming at mitigating flood problems and improving water availability in the dry season, particularly for agricultural purposes in northeast Thailand. However, the projects were done without assessment of water demand at the basin scale, and are operated by various government agencies mostly concern with individual project implementation, leaving the concern about river basin management has been a challenge (Floch, Molle, & Loiskandl, 2007). To date, while thousands of water projects were invested, they, however, have not performed as intended, and the flood and drought problems have still existed in the basin with more frequently.

To fill those gaps, in this paper, the flood and drought conditions that affect rice growth and production are summarized, and from which the total excess rainwater supply and water demand with respect to rice water needs at the basin scale are examined. We collect the data on current and ongoing water resources development projects and discuss the flaws and factors that influence the achievement of the projects. Finally, the practicable water management measures, along with corresponding responsible parties, to cope with both problems and to improve rice production, are proposed. They are selected based on the specific criteria emphasizing on flexible storage options covering both surface and subsurface storages, less expensive, ease of accessibility, and low environmental impacts to ensure the feasibility of the projects. The selected study area is the Mun River Basin in northeast Thailand, where the dual flood and drought problems dominate. To our knowledge, this is the first time of the analysis of all dimensions regarding floods and droughts in this area are assessed at the basin scale.

6.2 Data and methods

The flood hazards at different flood frequency at the Mun River Basin was assessed using the integrated hydrologic (SWAT) and hydraulic (HEC-RAS) models. The model development and calibration were undertaken in the previous study of Prabnakorn, Suryadi, Chongwilaikasem, and de Fraiture (2019) and was further employed in this study. The recurrence periods of 10, 25, 30, 50, and 100 years were derived from rainfall frequency analysis as they are usually adopted in the design of structural measures in Thailand. The data utilized in the study consist of precipitation time series from 1985-2015, land use, soil type, water level, discharge, etc., which were gathered from various ministerial departments in Thailand such as the Royal Irrigation Department (RID), the Meteorological Department, and the Land Development Department.

Drought, in this study, is defined based on the agronomist's view that is a condition of water stress that affects crop growth and yield (Pereira, Cordery, & Iacovides, 2002), and it was assessed on the basis of an imbalance between water supply and water demand in respect of rice growth at each growing month (Prabnakorn, Maskey, Suryadi, & de Fraiture, 2019). The water supplies are mean areal precipitation representing the volume of rainfall falling over the entire basin. The estimation was performed for each rice-growing month, over the growing season, and annual rainfall from all 53 rainfall-gauging stations by using Thiessen Polygons. The calculation of water demand, rice water needs, was carried out based on Brouwer, Prins, and Heibloem (1989), which includes the water needs for soil saturation, evapotranspiration, percolation and seepage losses, and the establishment of a water layer. The data used are more specific to the field (the study area), and the rice cultivars grown in the area. The crop factor (K_c) for the KDML105 and RD6, and the reference evapotranspiration (ET_0) were obtained from the studies and experiments by the RID. The percolation loss from paddy fields and the water needed for land preparation at the Mun River Basin are based on field observations. The values of the former vary from 1-3 mm/day, and that of the latter is about 200-250 mm (Kirdpitugsa & Kayankarnnavy, 2009), which in our case, the averages were used in the analysis. Moreover, the amounts of water needed for the establishment of water layers were obtained from Brouwer et al. (1989).

Further, the net water requirements for the total rice areas, 38,565 million m^2, over the whole basin were subsequently determined, which are the differences between mean areal precipitation and water needs for rice cultivation at all growing months (July-November) and over the growing season. The calculation was carried out on the assumption of equal land areas of the KDML105 and RD6.

The data of existing mitigation measures implemented in the area, i.e., the number of large-, medium-, and small-scale projects, and electric pumping stations, their storage capacity, and actual irrigated areas are not consistent between different government departments. The information from the RID was adopted in this study because it is the largest department responsible for water development projects in Thailand.

6.3 Existing and future flood and drought mitigation projects

During the last half century, thousands of water resources development projects have been implemented over the Mun River Basin (Table 6-1 and Figure 6-1). They are categorized with respect to RID's classification into three groups: large-, medium-, and small-scale. The large-scale projects have storage capacity ≥ 100 million m^3, or water surface area ≥ 15 km^2, or command area > 12,800 ha (Hydro and Agro Informatics

Institute, 2012). All large-scale projects are multi-purpose; 10 out of the total 12 projects are under the administration of the RID. The other two hydroelectric dams: Pak Mun Dam is under the Electricity Generating Authority of Thailand (EGAT), and Sirindhorn Dam is under the RID and EGAT.

Table 6-1 Summary of all water development projects in the Mun River Basin, 2016.

No.	Project	Number of projects	Storage (million m^3)	Irrigated Area (ha)
1	Large	12	3,367.54	156,756
2	Medium	194	1,270.33	172,986
3	Small	2,476	460.34	213,515
4	Electric pump	274	0.00	59,634
	Total	2,956	5,098.21	602,891

The medium-scale projects being implemented and complemented the large-scale schemes have storage capacity between 2 and 100 million m^3, or water surface area < 15 km^2, or command area between 480 and 12,800 ha, and, lastly, the small-scale projects are ones with construction period less than 1 year and no land compensation schemes embedded into development plans. Both are single-purpose projects primarily for domestic water use or irrigation, and very few projects for forest conservation. The vast majority of medium-scale projects are under the responsibility of the RID, and the rest are under the Department of Water Resources (Hydro and Agro Informatics Institute, 2012), while about 16 government agencies involve in small-scale projects (Patamatamkul, 2001). The important features of medium- and small-scale projects are present in Table 6-2.

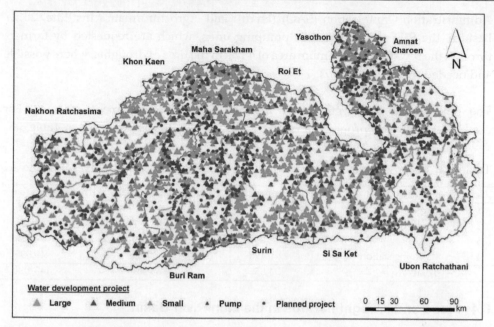

Figure 6-1 Water development projects at large, medium, and small scales, including electric pump, 2016; Source: The Royal Irrigation Department, Thailand.

Table 6-2 Summary of medium and small water development projects, including electric pumps in the Mun River Basin, 2016.

Project	Irrigation			Others		Total	
	No	Storage (million m³)	Irrigated area (ha)	No	Storage (million m³)	No	Storage (million m³)
Medium							
- Reservoir	112	1,059.55	118,777	22	70.65	134	1,130.20
- Weir	41	107.07	52,608	17	18.66	58	125.73
- etc.	1	0.00	1,600	1	14.40	2	14.40
Subtotal	154	1,166.62	172,986	40	103.71	194	1,270.33
Small							
- Reservoir	1,305	376.74	108,378	203	41.16	1,508	417.90
- Weir	827	39.32	102,777	100	0.67	927	39.99
- etc.	31	2.06	2,361	10	0.39	41	2.45
Subtotal	2,163	418.12	213,515	313	42.22	2,476	460.34
Pumping station	260	0.00	59,634	14	0.00	274	0
Total	2,577	1,585	446,135	367	145.93	2,944	1,730.67

Source: Royal Irrigation Department, 2016

Electric pumping stations (without storages) have been installed adjacent to the main river and its major tributaries throughout the basin (Table 6-1 and Figure 6-1). The pump operation and maintenance are under the responsibility of the Subdistrict

Administration Organization (SAO) (Hydro and Agro Informatics Institute, 2012). Besides, the RID provides mobile pumping units, which are requested by farmers through the SAO for the minimum area of 48 ha in the time of droughts, where possible and needed (Floch et al., 2007).

The government has continuously invested further development schemes. For example, the projects proposed and undertaken by the RID as presented in Table 6-3.

Table 6-3   Proposed and ongoing projects under the RID at the Mun River Basin, 2016.

No.	Project	Number of projects	Capacity (million m³)	Irrigable area (ha)
1	Large	3	135.37	0
2	Medium	148	937.97	46,340
3	Small	1,209	1,543.89	77,720
4	Pumping station	196	17.38	19,098
	Total	1,556	2,634.61	143,158

6.4   Flood and drought hazards in the Mun River Basin

6.4.1   Flood

The flood map for 30-year return period is presented, as an example, in Figure 6-2 (top). The extent of flooding is larger on the left bank than the right, and the flood depths vary mostly from 0 to 4 m. Approximately 60% of floodplain inundation is less than 1 m, which mostly occur at the upstream and central parts of the Mun River. At the river downstream, the extent of flooding is not as large as the upstream, but deeper. Moreover, due to the flat terrain, the duration of flooding is rather long, which can cause damage to crop growth and yields, including cities located in the flood-prone areas.

Figure 6-2 (bottom) entails the comparison of inundated areas of different scenarios, and their relevant information about affected areas and flooding volumes are demonstrated in Table 6-4. When the recurrence intervals increase, for example from 1 in 10 years to 1 in 25 years and so on, the increment of flooded areas are mostly observed at the central part of the basin because that area is flat with mild slopes, which locally called *Boong Taam*. It is the freshwater swamp forest, seasonally-flooded, and is the most important wetland in the northeast. Some parts of it are used for rice cultivation (Chusakun, 2013), and this rice is most affected by flooding. The flooding volume represents the excess water on the floodplains at different severity. The value is essential for effective planning and design of flood mitigation measures such as reservoirs or retention storages.

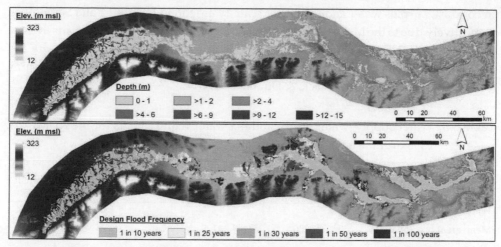

Figure 6-2 The flood map for 30-year return period (top), and comparison of inundated areas at 10-, 25-, 30-, 50-, and 100-year return periods (bottom) in the Mun River Basin, Thailand.

Table 6-4   Flooded area, flooded rice field, and flooding volume of different scenarios.

Return period	Flooding Area	Flooded rice fields		Flooding volume
(years)	(km²)	(km²)	(% total flooded area)	(million m³)
10	2,102	1,215	56%	3,061
25	2,535	1,521	60%	4,119
30	2,613	1,588	60%	4,371
50	2,873	1,791	62%	4,967
100	3,279	2,096	65%	5,962

6.4.2 Drought

Table 6-5 entails water supply, water needs for rice, and net water requirements of the wet-season rice at all growing months (July-November), and over the growing season. The amount of annual precipitation sufficiently fulfills the rice water needs for both varieties, but that of over the growing season does not. A little more rainwater is demanded for the KDML105 than the RD6. The maximum amount of water is required in July to saturate the soil for land preparation. Thus, droughts in this month do not harm rice but delay the onset of the growing season. The most severe water shortage appears in October in which rice is most vulnerable because two crucial stages: flowering and grain-filling, develop in this month. Both stages need much water to maintain the water layer throughout the month; otherwise, it will result in considerable yield reduction (Bouman, Lampayan, & Tuong, 2007; Fukai, Basnayake, & Cooper, 2000).

In contrast, surplus water and less water scarcity are found in August and September, respectively due to the highest precipitation and moderate water requirements in these two months. In November, although water deficit is observed, it does not create any adverse effects; instead, it will homogenize maturation and facilitate harvesting (Wopereis, Defoer, Idinoba, Diack, & Dugué, 2008).

Table 6-5   Water supply, water needs, and net water requirements for the KDML105, RD6 over the entire basin.

	Jul	Aug	Sep	Oct	Nov	Total (growing season)	Annual
Water supply to rice (mm)							
Mean areal precipitation	200	230	254	120	24	828	1,292
Water needs for rice (mm)							
KDML105	329	212	273	328	90	1,232	
RD6	329	235	266	300	85	1,215	
Average	329	223	270	314	87	1,224	
Net water requirements (million m³)							
KDML105	-2,485	358	-367	-4,022	-1,271	-7,787	1,166
RD6	-2,485	-94	-243	-3,479	-1,168	-7,469	1,483
Total	-4,970	264	-610	-7,501	-2,439	-15,256	2,649

6.5 Discussion on existing flood and drought mitigation projects

Although the government has implemented a tremendous number of water-related projects across the basin, the achievement is still far from the goals. The total storage capacity of all existing and ongoing projects (7,733 million m³) is sufficient to deal with the flooding volume at 100-year return period (5,962 million m³), and the most considerable water shortage for rice in October (-7,501 million m³). Consequently, flood and drought damage should not arise in the basin; however, in reality, the incidences of both events have been reported frequently, for example, they both occurred, in different periods, in years 2013 – 2016 (Department of Disaster Prevention and Mitigation, 2016; Hydro - Informatics Institute, 2019). The causes that expected benefits from the projects have never materialized are:

The geological and physical characteristics of the basin are the dominant constraints on water resources development and its achievement. The undulating topography and sandy soil with a low water-holding capacity limit viable sites for storage construction as it can be seen from the large projects that mostly located in the bottom part of the basin (Figure 6-1). To complete the projects, extensive resettlement involved making the projects difficult and costly tasks. Moreover, the basin lies over a rock salt formation; thus, salinity here, both in soils and water, is in part a natural phenomenon.

Together with high evaporation rates, above 30°C in the dry season, making irrigation water highly saline which nullifies irrigation benefits. The use of irrigation water on the sandy, infertile and saline soils, which commonly found in the area, significantly reduces rice yields and adds greatly to a salt burden of the land (Floch et al., 2007; Kamkongsak & Law, 2001; Shannon, 2005).

The justifications put forward for most water development projects in northeast Thailand are mainly from political reasons, rather than technical or economic rationales. The development projects have been served as a mean of politicians to gain electoral support in rural areas (Floch et al., 2007), thus, numerous times, the development projects were approved and built without sufficient studies. For example, Rasi Salai Dam was constructed without completed studies of environmental impacts, the suitability of soil or water demand (Matthews, 2011; Shannon, 2005). It is the Mun mainstream dam with the storage of 74.43 million m^3; its storage areas cover extensive low-lying land adjacent to the river called *Boong Taam*. It is a freshwater swamp wetland and forest from which local communities gathered forest products, fish and animal protein as food and traditional medicines. Some parts of the land are for agriculture - mostly rice because of fertile soils formed by nutrient-rich deposits during the wet season. The rice here could gain the average yield at 4.3 ton/ha higher than that in other areas, which is at 2.3-2.4 ton/ha. Moreover, during the dry season, parts of the forest that were covered with salts were used by villagers for their own consumption, trading with rice and selling. However, after the construction completed, the dam inundated and destroyed forest and floodplain ecosystems and blocked fish migration route by which local villagers lost their farmland (approximately 12,800 ha (Royal Thai Government, 2019)), saltpan farms, and food security and livelihoods.

To make matters worse, the stored water is highly saline, useless for irrigation, and salination has spread into agricultural land, causing damage to rice and vegetables. (Chusakun, 2013; Kamkongsak & Law, 2001; Matthews, 2011; Thai Baan researchers, 2005). This case provides an example of social and environmental devastation which finally leads to vast amount of compensation, 2.5 billion Baht (US$0.08 billion) (from 1997 to 2019) (BangkokBizNews, 2019; The Nation, 2019) almost triple of the construction cost at about 872 million Baht (US$27.8 million) (Chusakun, 2013). The similar cases have also been reported at Pak Mun and Hua Na Dams (also in the Mun River Basin) and in other projects, which resulted from political motivation with poorly plan, inadequate research, and lack of transparency and public participation.

The construction of storage is not paralleled by a linear expansion of irrigation service areas along with lacks of regular inspection and maintenance programs leading most of the schemes do not function properly. For example, the Lam Se Bai Weir- the large-scale project, only one distribution system, from the whole 3 systems, was completed 9 years behind schedule. The completed system was initially designed for a total irrigable area of 13,600 ha, but, in reality, the irrigable area is only 3,168 ha (23% of the target), and from the 2014 report the service area is only 320 ha in the wet season and 112 ha in the dry season. Moreover, many parts of the systems (i.e., electric pumps, dikes, distributed canals, gates, etc.) fell into disrepair because of lack of maintenance, and some of them could not be used (State Audit Office of the Kingdom of Thailand, 2016). These similar situations also happen for many other large-scale projects, and most of the small-scale projects have no distribution systems, and in practice, the irrigated areas may reduce by 30-40% (Hydro and Agro Informatics Institute, 2012). The operation and maintenance procedures of the small-scale projects have been left to local people with little explanation and assistance from the related agencies. As a result, many of the systems were incomplete, non-existent, or in poor condition (Floch et al., 2007; Patamatamkul, 2001).

The electric pumping schemes have limited success because of insufficient river flows in the dry season (Floch et al., 2007). The actual irrigated area served by one pump has been only 33.6 ha although the Department of Energy Development and Promotion (DEDP) claimed that each electric pump project is capable of irrigating an average of 240 ha (Kamkongsak & Law, 2001; Shannon, 2005).

Lack of a single-commanded authority and an integrated approach as the institutional framework is highly fragmented. There are 38 water-related agencies involved in water resources development (Office of the National Water Resources, 2018a) in Thailand. Most of them are mostly concerned with individual project implementation, lacking unity and coordination. This frequently leads to overlapped or overlooked service areas, inconsistent strategies, and budget allocation, including other impediments to efficient water management and project achievement (Netherlands Embassy in Bangkok, 2016).

Reservoir operation and management in the area is challenging due to the conflict over the operation. For flood control, water levels in the reservoirs need to be lowered as much as possible, whereas to handle drought, they need to keep water to their greatest extent possible. The operation is thus a key factor for either flood and drought mitigation or exacerbation. It, therefore, remains a major undertaking in the basin.

6.6 Institution framework

In October 2017, the Thai government established the Office of the National Water Resources (ONWR), a neutral agency, which is responsible for the integration of information, plans, projects, budgets, and administration associated with water resources development in Thailand. It is mandated to facilitate, advise, monitor, and evaluate the implementation of all water-related agencies country-wide (Figure 6-3) (Office of the National Water Resources, 2018a). Afterwards, the first National Water Resource Act, B.E. 2561 was enacted in December 2018 (Office of the National Water Resources, 2018b) upon which the 20-year water resources management master plan (2018-2037) drafted by the ONWR and other related agencies, was approved by the government in June 2019 (National News Bureau of Thailand, 2019). It is hoped that this reform and policy formulation will bring about the unity in water resources management in Thailand, and can effectively and sustainably tackle the chronic flood, drought, and wastewater problems in the country.

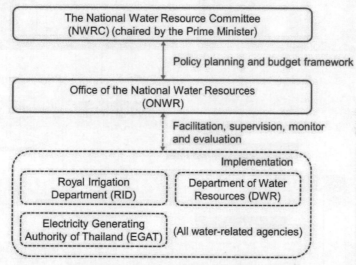

Figure 6-3 The governance structure of water resources development in Thailand.

6.7 Potential water management solutions

Water storage has a vital role in tackling floods and droughts to enhance crop production. There is a wide range of water storage options, as depicted in Figure 6-4. The best option is to focus on flexibility in storage systems, wherever possible combining a variety of water storage types that will provide a crucial mechanism for adaptation to the coming climate extremes (International Water Management Institute (IWMI), 2009). Therefore, we proposed the diverse storage infrastructure encompassing all options, except soil moisture (in the last column of Figure 6-4 and Table 6-6). Because of soil moisture conservation techniques, for example, bunding and terracing prevail commonly in rice fields since the past. The proposed solutions were selected based on the following criteria:

1) Address both floods and droughts,

2) Simple and affordable, building on existing initiatives,

3) Proved effective in previous studies or pilot projects,

4) Apply Nature-based solutions where appropriate.

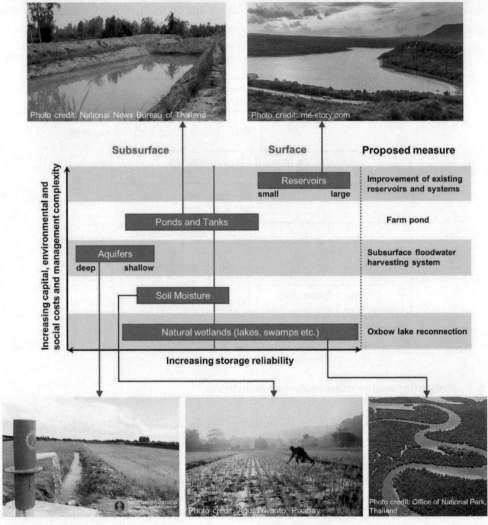

Figure 6-4 The continuum of water storage options (International Water Management Institute (IWMI), 2009), and our proposed measures (the last column) for solving floods and droughts at the Mun River Basin, Thailand.

Those criteria facilitate the selection of potential and reasonable measures feasible to be implemented to solve floods and droughts with minimizing environmental transformation and impacts. They need to be executed with careful attention by the cooperation among the corresponding responsible parties as presented in Table 6-6.

Table 6-6 Summary of proposed measures and strategies along with responsible parties.

Proposed measures and strategies	Responsibility		
	Central government	Local government	Farmers
Improvement of existing measures			
- Completion of designed distribution systems, and repair of the deteriorated schemes	✓	✓	✓
- Planning and execution of regular inspection and maintenance practices		✓	✓
- Post-implementation reviews of the controversial projects	✓	✓	✓
- Reservoir operation & early warning systems	✓	✓	✓
Individual farm pond		✓	✓
Subsurface floodwater harvesting system		✓	✓
Oxbow lake reconnection		✓	✓

6.7.1 Improvement of existing measures

Despite investment in new water development projects, what the government should concern now is the inefficiency and ineffectiveness of the existing measures, which is the fundamental problem here. Some recommendations to improve the performance, efficiency, and add value to the in-situ measures are given below.

The construction of designed distributed canals and systems, repair or rehabilitation of the deteriorated elements and schemes should be accomplished. Moreover, in some areas, the distribution canals may need to be reconsidered because of the salinity problem. The lessons from the irrigation scheme in Sisaket province document that the irrigation canals are used every 3-4 years to supplement the wet-season crops during dry years, and some canals are left unused (Shannon, 2005). This knowledge and experiences of local farmers should thus take into account together with the financial support from the central government and technical assistance from local government agencies.

Additionally, it is necessary to have a plan of routine inspection and maintenance in which the division of duties between the local government and farmers should be clearly described, and at the beginning, the farmers' responsibilities should be sufficiently supported by government measures or assistance. The local government should develop and implement the maintenance plan in collaboration with local farmers. Involvement of farmers is essential for maximizing returns from irrigation projects, and this should receive high priority (Jurriëns & Jain, 1993). Local water-user groups can devise institutions to manage the irrigation system sustainably as their livelihoods depend on that common property, they thus have the greatest incentives to maintain it over time (Bromley, 1992; Meinzen-Dick, Raju, & Gulati, 2002). To make farmers involvement effective, legislative backing and financial incentives will be required in the initial years (Jurriëns & Jain, 1993).

For the controversial projects such as Rasi Salai, Hua Na, and Pak Mun dams, if the government wishes to continue the plans, the post-implementation monitoring and reviews should be sufficiently conducted over many years, although they are not a legal obligation. It is because the large water projects, which resulted in ecosystem degradation, not only affect local people livelihoods but also connect to several health issues, e.g., snail-borne diseases such as schistosomiasis (blood fluke infection) or opisthorchiasis (liver fluke infection), and mosquito-borne diseases such as malaria and filiariasis (Keiser et al., 2005; Service, 1991; Shannon, 2005). These undesirable side-effects are usually received little attention because other priorities, e.g., maximizing or increasing agricultural production and economics, dominate the scene (Service, 1991). The reviews should include all relevant aspects, including environmental, social, health, economic impacts, and need the greater intersectoral and interdisciplinary collaboration of all stakeholders. The knowledge of local communities should not be ignored or undervalued as the locals are directly impacted by the development projects.

For reservoir operation, the government should implement operational rules in daily operations by a preference for up-to-date measurement and real-time forecasts of weather and flow conditions. These operational rules should include multi-purpose optimization real-time reservoir operation with model predictive control to support real-time operations. In addition, the results from reservoir operation and forecast can also be used for flood early warning at the downstream of the reservoir. Early warning is used to issue warnings when a flood is imminent or already occurring. Consequently, it helps to minimize human fatalities, injuries, and health risks as well as properties damage. However, all of these suggestions will not be successfully implemented if local stakeholders are not involved in the planning and decision process. It is because stakeholder involvement is likely to result not only in better information and knowledge at the local sites but also in understanding community resilience and adaptation capacity (Lang et al., 2012).

6.7.2 Individual farm pond

One of the main justification for the promotion of the investment in water-related projects in northeast Thailand was dry-season rice agriculture; however, the large-scale adoption of dry-season rice farming never happened, less than 5% across the basin (Floch et al., 2007). From our point of view, it is difficult for the large-scale dry-season rice cultivation to be possible if other factors, i.e., cropping pattern, field irrigation practices, etc. remain unchanged. It is because the total precipitation over the wet rice-growing season over the past 30 years have never met the rice water needs. Moreover, although the annual rainfall did at some rain-gauge stations, particularly at the stations on the east part of the basin, it is still not sufficient for both wet- and dry-season rice agriculture (Figure 6-5). Furthermore, many farmers are reluctant to plant the second rice because of low yields, fewer profits due to expenses of electricity (for irrigation pumps), chemical fertilizer, and pesticide. Also, the farmers in many areas do not use irrigated water because it is highly saline, and that significantly reduce yields (Kamkongsak & Law, 2001; Shannon, 2005).

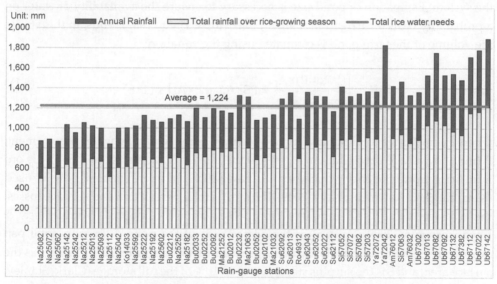

Figure 6-5 Averages of annual precipitation and total precipitation over the rice-growing season at all stations for 1984-2016, and the total water requirement for rice. Rain-gauge stations are listed from west to east.

Base on the above situations and as precipitation is still the primary source for agriculture, the only objective should be sufficient water for the wet-season rice cultivation. If rainwater is efficiently caught and stored, there would have been enough for the wet-season rice, and if there is surplus water, farmers can gain additional benefits from other second crops such as soybeans, watermelons, and vegetables suitable to the season which require less amount of water than the dry-season rice.

According to the 'New theory' agriculture developed by King Bhumibol Adulyadej of Thailand, after the large-, medium-, and small- reservoirs, distribution canals, and systems were constructed, villagers should build small farm ponds and connect to the constructed reservoirs in order to be able to efficiently manage the water and their farmland (The Chaipattana Foundation, n.d.). The pilot project was carried out at Ban Limthong, a village with 108 households in Buri Ram province in the Mun River Basin, where floods and droughts periodically occur. The results reveal that by adopting the 'New theory,' the flood and drought problems that plagued the community for decades are solved, and it also created a stable buffer for the village (van Steenbergen, Tuinhof, Knoop, & Kauffman, 2011).

As the reservoirs have been already constructed over the entire basin, we, therefore, propose to build individual farm ponds, another imperative component to complete the system. Estimation of pond size is necessary to ensure the availability of water on a probability basis for irrigation (Palmer, Barfield, & Haan, 1982). Thus, in this study, the volume of the farm pond is estimated using the equation:

$$P_v = \left[\frac{(WR - P) \times A_I}{E_f} + PS \times A_w + E \times A_s\right] \times (1+S) \qquad (6.1)$$

where P_v (m^3) is the total pond volume, WR (mm) is the total crop water requirement, P (mm) is average precipitation, A_I (m^2) is the total irrigation area, and E_f (%) is on-farm irrigation efficiency which, in this case, 80% for open channel flow is used (Kirdpitugsa & Kayankarnnavy, 2009). PS (mm) is total percolation and seepage losses; the average value of 2 mm/day from observation is adopted (Kirdpitugsa & Kayankarnnavy, 2009), A_w (m^2) is the wetted area of the pond, E (mm) is total pan evaporation obtained from Hydro and Agro Informatics Institute (2012), A_s (m^2) is the surface area, and S is the surplus storage, at least 10% of the full storage (Clark, Stanley, Zazueta, & Albregts, 2002). It is important to note that PS and E account for the total duration of the water to be stored in the pond before irrigation.

We estimated the pond size of the only month of October when it is a high-water shortage, and rice is most fragile to drought. It is because the precipitation here starts from mid-May and reaches the maxima in August or September when some areas are flooding because of extreme precipitation intensity. We thus keep this excess runoff, which generally causes damage to rice growth and yields (Prabnakorn et al., 2018) then utilizes it in October when precipitation declines. Thus, besides reserving water for rice cultivation, another advantage of the pond is increasing the total water storage capacity, which can reduce and prevent flooding.

The sizes of the pond are: 3,100 m^3 per 1 ha (500 m^3 per 1 rai) (rai is the local Thai measurement of areas = 1600 m^2) if only the KDML105 is planted; 2,670 m^3 per 1 ha (435 m^3 per 1 rai) if only the RD6 is planted; and 2,880 m^3 per 1 ha (470 m^3 per 1 rai) if both the KDML105 and RD6 are equally planted. The pond sizes are estimated relative to the average precipitation over the entire basin; thus, water deficits in some areas or some years may arise if rainfall is less than the average. However, if the full distribution systems are constructed, and the farm ponds are connected with medium and small reservoirs, these reservoirs will serve as water reserves for repletion of the ponds to ensure water availability.

Considering the pond depth at 3 m, the pond sizes occupy about 10% of the farm area, in other words, farmers have to sacrifice about 10% of their land to ensure water availability in October, which will significantly enhance rice yields. It is thus a trade-off between agricultural land and water security for rice cultivation. If farmers prefer the full capacity to be assured of no water deficits for rice, about 15% of the area is required for excavation of the ponds. Moreover, the pond sizes can be redesigned to be more cost-effective and suitable for each farm by using the above equations along with specific climatic conditions, cropping patterns, and soil properties in each area.

Farmers themselves are the key driver for farm pond development. However, due to their low income, they still need financial support, i.e., funding loans or subsidies from the government, foundations, or private sectors, to carry out the relevant activities. Also, cooperation with local government agencies for the construction of irrigation canal systems, data support and consulting (van Steenbergen et al., 2011).

6.7.3 The subsurface floodwater harvesting system

After a long time, that flood and drought mitigation measures in Thailand have solely revolved around surface water storage; however, recently, Managed Aquifer Recharge (MAR) has been recognized. Pavelic et al. (2012) apply the MAR concept to capture floodwater to recharge subsurface storage, and from which the water is drawn as groundwater for agriculture uses during the dry season. The pilot trial is in the Lower Yom River Sub-basin of the Chao Phraya River Basin in Thailand, and the findings reveal that the groundwater recharge reduced the magnitude of flooding and generated approximately USD 250 M/year in farm earnings from dry season production of irrigated rice.

This approach is an element of water storage that should not be neglected (International Water Management Institute (IWMI), 2009). It can be adopted to mitigate flood and drought problems in the Mun River Basin although some parts of the basin, especially the Mun River upstream and the tributaries in the upper part of the basin, have salinity problem (total dissolved solids: TDS > 1500 mg/l) (Figure 2-8). It is not worth the investment to conduct the MAR in those areas because thought it can reduce flood peaks, no recovery benefit for productive use is obtained because the groundwater with the TDS ≥ 2000 mg/l is not suitable for irrigation, and it will cause severe yield reduction (Ayers & Westcot, 1985). We, therefore, propose to implement the MAR at the tributaries upstream at the lower and the eastern parts of the basin where there are fewer or no salinity hazards. This will result in less flood volume, and its adverse impacts downstream, and the recharge water is vital for supplementing surface water utilization, and to ensure ongoing groundwater exploitation by more than 75% of villages over northeast Thailand (Srisuk, Sriboonlue, Buaphan, & Hovijitra, 2001) for agriculture and consumption.

Though the MAR has the advantage of a smaller footprint on the landscape, the total cost of the whole system establishment, operation, and maintenance are too high for the farmers alone. The support from the local government agencies is necessary in terms of budget availability and technical and administrative work. A thorough study regarding the MAR implementation in the basin is inevitable to ensure that only floodwater is captured without significantly impacting the supply-demand balance downstream. This requires a close partnership between the local government and farmers. The farmers also play an important role in the ongoing operational performance and maintenance of the flood harvesting structure. Financial or incentives maybe need to soliciting their efforts and continuing participation and their contribution to reducing the magnitude of flooding downstream (Pavelic et al., 2012).

6.7.4 Oxbow lake reconnection

Reconnection of oxbow lake is a Nature-Based Solutions measure that can help in slowing run-off and reducing flood peaks as well as storing excess water in the lake to use during dry spells. Additionally, the oxbow lakes accommodate habitat diversity as spawning places for fish and other aquatic groups (Obolewski & Glińska-Lewczuk, 2011). There is substantial evidence of fruitful implementation of oxbow reconnection in many places, for example at the Moravá River in Slovak (CIS, 2006), and Tisza River in Hungary (Fisher & Stratford, 2008); the oxbow lake reconnections prove successful in increasing flood protection capacity and restoration of hydrological connectivity, and significantly benefiting ecological status. In order to reconstruct oxbow lake, the fundamental elements of the measure should include an inlet work at the entrance and an outlet structure to control the floodwater as in Figure 6-6.

Figure 6-6 Components of reconnecting oxbow lakes.

Regarding the physical characteristics of the Mun River that is meandering and anabranching, and as a result of the hydrological and geomorphological process, numerous oxbow lakes have been cut-off from the main river causing a reduction in river's storage capacity. Thus, reconnecting the oxbow lakes is another effective way for flood and drought mitigation. The approach makes use of the natural river characteristic; therefore, the relevant work is not as vast as the new whole projects, and likely to be done by the cooperation between villagers and local government agencies. The local people also have a major part in operation and maintenance with financial and incentive support as well as technical assistance from the local water agencies.

Besides the proposed measures mentioned above, the non-structural measures such as field management practices, land use planning and policy should be taken into consideration. The field management techniques for rice cultivation such as furrow irrigation, alternate wetting and drying irrigation (AWD), rice ratooning, diminish water demand for rice cultivation and increase water productivity. The accomplishment of those practices in many countries is documented in many studies, for example, Atta (2008); Lampayan et al. (2004); Santos, Fageria, and Prabhu (2003); Wang et al. (2018). Whereas land-use planning minimizes development in flood-prone areas, and conserve floodplains and wetlands as natural water storages. This will reduce flood and drought damage, and is essential for ecosystems.

6.8 Conclusions

The study aims to solve flood and drought damage to rice cultivation in the Mun River Basin in Thailand; therefore, all dimensions related to both hazards are considered at the basin scale, including affected areas, flooding volumes, water supply, water demand, water deficits, etc. Subsequently, the coping capacity of the current and ongoing projects is evaluated, and the flaws and factors influencing their achievement are discussed. Then, we propose the potential measures along with corresponding responsible parties to cope with both problems and to enhance rice yields.

The findings present that the total storage capacity of all existing measures is sufficient to tackle floods and droughts. However, the opposite often proves to be the case in reality – both hazards still occur in different periods and more frequently. The major causes of this failure involve geological and physical characteristic constraints, political motivation without technical and economic support in project development, incomplete construction of irrigation distribution systems and lack of regular inspection and maintenance, the challenge of reservoir operation, and fragmented institution framework. However, recently, the government reformed the institution structure and launched the first National Water Resource Act and the 20-year master plan with the expectation of unity and consistency in water resources management in Thailand.

We propose the potential measures and suggestions to solve flood and drought problems focusing on flexibility in storage systems by combining a wide range of water storage options: improvement of existing projects, farm ponds, subsurface floodwater harvesting system, and oxbow lake reconnection. They are selected base on four criteria: addressing both floods and droughts, simple and economical, proved effective in previous studies, and apply Nature-based solutions where appropriate. The criteria facilitate the selection of potential and reasonable measures feasible to be implemented with minimizing environmental transformation and impacts.

The paper provides an understanding of flood and drought impacts on rice cultivation, the coping capacity, and weaknesses of the in-situ mitigation projects at the basin scale. These facts are vital for all stakeholders to realize the actual circumstance in order to search for proper solutions. For the proposed measures, if they are cautiously executed with appropriate pre- and post-project studies and reviews, the flood and drought problems in the area may be reduced or solved sustainably. Moreover, the sizes of farm ponds can be recalculated to be more specific to each area or applied in other basins with single- or multi-crops besides rice, or to cover water uses all year round by adjusting the parameters in the offered equation.

References

Artlert, K., & Chaleeraktrakoon, C. (2013). Modeling and analysis of rainfall processes in the context of climate change for Mekong, Chi, and Mun River Basins (Thailand). *Journal of Hydro-environment Research, 7*(1), 2-17.

Atta, Y. I. (2008). *Innovative Method for Rice Irrigation with High Potential of Water Saving.* Paper presented at the ICID Congress, integrated water resources management–from concepts to actions. , Lahore, Pakistan.

Ayers, R., & Westcot, D. (1985). Water quality for agriculture. FAO Irrigation and drainage paper 29 Rev. 1. *Food and Agricultural Organization. Rome.*

BangkokBizNews (Producer). (2019). Cabinet agrees Bt600m for Rasi Salai Dam compensation (in Thai). *BangkokBizNews.* Retrieved from https://www.bangkokbiznews.com/news/detail/832061

Bates, B. C., Kundzewicz, Z. W., Wu, S., & Palutikof, J. (2008). Climate change and water. IPCC Technical paper VI. In *Climate change and water. IPCC Technical paper VI:* IPCC Secretariet.

Bouman, B. A. M., Lampayan, R. M., & Tuong, T. P. (2007). *Water management in irrigated rice: coping with water scarcity.* Los Banos, Philippines: International Rice Research Institute.

Bromley, D. W. (1992). *Making the commons work: Theory, practice and policy* (D. W. Bromley Ed.). San Francisco, California: Institute for Contemporary Studies Press.

Brouwer, C., Prins, K., & Heibloem, M. (1989). Irrigation Water Management: Irrigation Scheduling. In *Irrigation water management: Training manual no. 4.* Rome, Italy: Food and Agriculture Organization of the United Nations (FAO).

Centre for Research on the Epidemiology of Disaster (CRED). (2015). EM-DAT: The International Disaster Database. Retrieved January 21, 2015, from Centre for Research on the Epidemiology of Disaster - CRED, Université Catholique de Louvain, Brussels, Belgium www.emdat.be

Chusakun, S. (2013, March 2013). The truth of Rasi Salai: "Paa Taam" and the dam of the Kong-Chi-Mun project (in Thai). *Thang E-Shann, 11.*

CIS. (2006). *Case Studies: Potentially relevant to the improvement of ecological status/ potential by restoration/mitigation measures (Separate Document of Water Framework Directive Technical Report).* Retrieved from

Clark, G., Stanley, C., Zazueta, F., & Albregts, E. (2002). Farm ponds in Florida irrigation systems. *Extension Bulletin, 257.*

Department of Disaster Prevention and Mitigation. (2016). *Disaster event report 2016 (in Thai).* Retrieved from http://www.disaster.go.th/th/index.php

Department of Water Resources of Thailand. (2016). *Summary of the results of drought prevention and mitigation year 2015-2016. Final Report (in Thai).* Bangkok, Thailand.

Fisher, J., & Stratford, C. (2008). Does reconnection mean restoration for an oxbow lake, Hungary? *International journal of river basin management, 6*(3), 201-211.

Floch, P., Molle, F., & Loiskandl, W. (2007). Marshalling water resources: a chronology of irrigation development in the Chi-Mun River Basin, Northeast Thailand. *Colombo, Sri Lanka: CGIAR Challenge Program on Water and Food.*

Fukai, S., Basnayake, J., & Cooper, M. (2000). Modelling water availability, crop growth, and yield of rainfed lowland rice genotypes in northeast Thailand. *Characterising and understanding rainfed environments. Los Baños, Philippines, IRRI,* 111-130.

Hydro - Informatics Institute. (2019). Water events records (in Thai). Retrieved from http://www.thaiwater.net/v3/archive

Hydro and Agro Informatics Institute. (2012). Data collection and analysis for developement of data inventory of 25 basins in Thailand: the Mun River Basin. Final Report (in Thai), Bangkok, Thailand.

International Water Management Institute (IWMI). (2009). Flexible water storage options and adaptation to climate change. *IWMI Water Policy Brief*(31), 5. doi:http://dx.doi.org/10.3910/2009.412

Johnston, R., Lacombe, G., Hoanh, C. T., Noble, A., Pavelic, P., Smakhtin, V., . . . Choo, P. S. (2010). *Climate change, water and agriculture in the Greater Mekong Subregion* (Vol. 136): IWMI.

Jurriëns, M., & Jain, K. (1993). Maintenance of irrigation and drainage systems. *International Institute for Land Reclamation and Improvement, Wageningen, the Netherlands.*

Kamkongsak, L., & Law, M. (2001). Laying waste to the land: Thailand's Khong-Chi-Mun irrigation project. *Watershed, 6*(3), 25-35.

Keiser, J., de Castro, M. C., Maltese, M. F., Bos, R., Tanner, M., Singer, B. H., & Utzinger, J. (2005). Effect of irrigation and large dams on the burden of malaria on a global and regional scale. *The American journal of tropical medicine and hygiene, 72*(4), 392-406.

Kirdpitugsa, C., & Kayankarnnavy, C. (2009). Chi Basin Water Uses System Study by Developed Models. *Kasetsart Engineering Journal, 4*(62), 63-77.

Lampayan, R., Bouman, B. A., De Dios, J., Lactaoen, A., Espiritu, A., Norte, T., . . . Soriano, J. (2004). *Adoption of water saving technologies in rice production in the Philippines*: Food & Fertilizer Technology Center.

Lang, D. J., Wiek, A., Bergmann, M., Stauffacher, M., Martens, P., Moll, P., . . . Thomas, C. J. (2012). Transdisciplinary research in sustainability science: practice, principles, and challenges. *Sustainability Science, 7*(1), 25-43.

Manomaiphiboon, K., Octaviani, M., Torsri, K., & Towprayoon, S. (2013). Projected changes in means and extremes of temperature and precipitation over Thailand under three future emissions scenarios. *Climate Research, 58*(2), 97-115.

Matthews, N. (2011). Rasi Salai, Thailand. In B. R. Johnston, L. Hiwasaki, I. J. Klaver, A. R. Castillo, & V. Strang (Eds.), *Water, cultural diversity, and global environmental change: Emerging trends, sustainable futures?* : Springer Science & Business Media.

Meinzen-Dick, R., Raju, K. V., & Gulati, A. (2002). What affects organization and collective action for managing resources? Evidence from canal irrigation systems in India. *World Development, 30*(4), 649-666.

National News Bureau of Thailand. (2019, 21 Jun. 2019). The Cabinet of Thailand approves a 20-year plan on water resources management (in Thai). Retrieved from http://nwnt.prd.go.th/centerweb/news/NewsDetail?NT01_NewsID=TCAT G190620183846722

Netherlands Embassy in Bangkok. (2016, 18 May 2016). The Water Sector in Thailand. Retrieved from https://www.netherlandsworldwide.nl/documents/publications/2016/05/1 8/factsheet-water-sector-in-thailand

Obolewski, K., & Glińska-Lewczuk, K. (2011). Effects of oxbow reconnection based on the distribution and structure of benthic macroinvertebrates. *Clean–soil, air, water, 39*(9), 853-862.

Office of Agricultural Economics. (2018). *Agricultural Statistics of Thailand 2018.* Annual Report (in Thai), Bangkok, Thailand.

Office of the National Water Resources. (2018a). *The first chapter of national water resources.* Bangkok, Thailand.

Office of the National Water Resources. (2018b). *The National Water Resource Act, B.E. 2561 (in Thai).* Bangkok, Thailand Retrieved from http://www.onwr.go.th/?page_id=4184.

Palmer, W., Barfield, B., & Haan, C. (1982). Sizing farm reservoirs for supplemental irrigation of corn. Part I: Modeling reservoir size yield relationships. *Transactions of the ASAE, 25*(2), 372-0376.

Patamatamkul, S. (2001). *Development and management of water resources in the Korat Basin of northeast Thailand.* Paper presented at the Development and management of water resources in the Korat Basin of northeast Thailand, Manila, Philippines.

Pavelic, P., Srisuk, K., Saraphirom, P., Nadee, S., Pholkern, K., Chusanathas, S., . . . Smakhtin, V. (2012). Balancing-out floods and droughts: opportunities to utilize floodwater harvesting and groundwater storage for agricultural development in Thailand. *Journal of Hydrology, 470,* 55-64.

Pereira, L. S., Cordery, I., & Iacovides, I. (2002). *Coping with water scarcity:* Springer.

Prabnakorn, S., Maskey, S., Suryadi, F., & de Fraiture, C. (2016, 6-8 Nov 2016). *Climate and Drought Trends and Their Relationships with Rice Production in the Mun River*

Basin, Thailand. Paper presented at the International Commission on Irrigation and Drainage (ICID) Conference Proceedings, Chiang Mai, Thailand.

Prabnakorn, S., Maskey, S., Suryadi, F., & de Fraiture, C. (2018). Rice yield in response to climate trends and drought index in the Mun River Basin, Thailand. *Science of the Total Environment, 621*, 108-119. doi:10.1016/j.scitotenv.2017.11.136

Prabnakorn, S., Maskey, S., Suryadi, F., & de Fraiture, C. (2019). Assessment of drought hazard, exposure, vulnerability, and risk for rice cultivation in the Mun River Basin in Thailand. *Natural Hazards, 97*, 891-911. doi:10.1007/s11069-019-03681-6

Prabnakorn, S., Suryadi, F., Chongwilaikasem, J., & de Fraiture, C. (2019). Development of an integrated flood hazard assessment model for a complex river system: a case study of the Mun River Basin, Thailand. *Modeling Earth Systems and Environment.* doi:10.1007/s40808-019-00634-7

Royal Thai Government. (2019, May 2, 2019). Pay compensation to people affected by the Rasi Salai dam project (in Thai). Retrieved from https://www.thaigov.go.th/news/contents/details/20369

Santos, A., Fageria, N., & Prabhu, A. (2003). Rice ratooning management practices for higher yields. *Communications in soil science and plant analysis, 34*(5-6), 881-918. doi: 10.1081/CSS-120018981

Service, M. (1991). Agricultural development and arthropod-borne diseases: a review. *Revista de saúde pública, 25*, 165-178.

Shannon, K. L. (2005). The social and environmental impacts of the Hua Na dam and Khong-Chi-Mun project: The necessity for more research and public participation. *Presentation at Water for Mainland Southeast Asia, 30.*

Snidvongs, A., Choowaew, S., & Chinvanno, S. (2003). Impact of climate change on water and wetland resources in Mekong river basin: Directions for preparedness and action. *Change, 2*(2).

Srisuk, K., Sriboonlue, V., Buaphan, C., & Hovijitra, C. (2001). *The potential of water resources in the Korat Basin.* Paper presented at the Natural resource managment issues in the Korat basin of Northeast Thailand: An overview, Los Banos, Philippines.

State Audit Office of the Kingdom of Thailand. (2016). *Inspection Report of The Lam Se Bai Weir Project (in Thai).* Bangkok, Thailand: State Audit Office of the Kingdom of Thailand.

Thai Baan researchers. (2005). *Rasi Salai Thai Baan Research (in Thai).* Bangkok, Thailand: Southeast Asia Rivers Network.

The Chaipattana Foundation. (n.d.). Sufficiency Economy & New Theory. Retrieved from http://www.chaipat.or.th/eng/concepts-theories/sufficiency-economy-new-theory.html

The Nation (Producer). (2019). Cabinet sets aside over Bt500m for last of Rasa Salai Dam victims. *The Nation.* Retrieved from https://www.nationmultimedia.com/detail/national/30367476

van Steenbergen, F., Tuinhof, A., Knoop, L., & Kauffman, J. (2011). *Transforming landscapes, transforming lives: the business of sustainable water buffer management*: 3R Water.

Wang, Z., Gu, D., Beebout, S. S., Zhang, H., Liu, L., Yang, J., & Zhang, J. (2018). Effect of irrigation regime on grain yield, water productivity, and methane emissions in dry direct-seeded rice grown in raised beds with wheat straw incorporation. *The Crop Journal, 6*(5), 495-508.

Wopereis, M., Defoer, T., Idinoba, P., Diack, S., & Dugué, M. (2008). Participatory learning and action research (PLAR) for integrated rice management (IRM) in inland valleys of sub-Saharan Africa: technical manual. *WARDA Training Series. Africa Rice Center, Cotonou, Benin, 128*, 26-32.

7

Synthesis and main contributions & prospects of further research

7.1 Synthesis

Floods and droughts are climate extremes that account for more than 80% of people affected by natural disasters worldwide (Table 1-2). Over the period from 1900-2014, floods killed approximately 7 million people, affected the lives of 3,500 million people, and cost USD 640 billion in damages. The corresponding figures for droughts were: 12 million killed, 2,170 million people affected, and USD 133 billion in damages. Geographically, the majority of people affected by both hazards are in Asia (95% from flooding, and 80% from droughts (CRED, 2014).

Agriculture, which occupies a large proportion of landscape, is usually impacted by both calamities. The damage typically involves crop production losses (OECD, 2016). This puts pressure on food security and can lead to famine and poverty. The poor are the most vulnerable and suffer significantly in terms of loss of lives and livelihood from adverse impacts of extreme climate events (ADPC-UNDP, 2005).

Consequently, considerable efforts have been devoted to solving flood and drought problems and building communities' resilience and coping capacities to the disasters. However, perspectives on and approaches to flood and drought mitigation are typically developed in a separate manner. This is due to differences in natural characteristics and times of occurrence of floods and droughts. As a result, the opportunities to use floodwater in times of water shortage (either by surface water storage or through groundwater recharge) may be overlooked (Pavelic et al., 2012; UNEP-IETC, 1998). Moreover, in some cases, overlap and inconsistency in operations and management may arise, which can be highly inefficient.

Floods and droughts co-exist in many river basins, although their occurrences are usually different in time and/or space (Pavelic et al., 2015). For this reason, and to fill those gaps mentioned above, this thesis proposes to adopt a holistic approach instead of the traditional approach. The holistic approach attempts to tackle and assess both hazards at the same time by employing integrated measures and strategies. The study is carried out at the Mun River Basin in Thailand where rice cultivation dominates. Accordingly, the main objective is *to assess flood and drought impacts on rice cultivation at the basin scale and attempt to tackle them simultaneously by using integrated measures and strategies.* Based on that, the primary research question is formulated as '*What are the extents of flood and drought impacts on rice cultivation in the basin and how to tackle them simultaneously?*'

In this chapter, all the findings in the thesis are synthesized and discussed in broad perspectives in relation to technical methods used for flood and drought analysis; flood and drought impacts on rice cultivation and yields; and mitigation measures and policy development. Lastly, the chapter closes with a brief recommendation for further studies.

7.1.1 Flood and drought analysis

Floods and droughts are absolutely distinct in terms of their characteristics and occurrences. Flooding is apparent in terms of its onset, and intensity. Droughts, on the other hand, have a slow-onset nature and cumulative, nonstructural and nonobvious impacts, meaning their severity is more difficult to quantify than floods (Wilhite & Vanyarkho, 2000). For rice crops, a flood event can be considered to occur the moment the land is inundated until the floodwaters recede and can be measured in terms of extent or depth. For a drought, however, the start and end times are not clear, nor is the extent or even the severity of the event.

As a consequence, approaches for the identification of vulnerable areas to floods and droughts are somewhat mutually exclusive. For flooding, flood model simulation (1D, quasi2D, 2D, coupling 1D2D) is one of the most acceptable and reliable methods to schematize and predict flood extents and severity. Droughts are mostly estimated using drought indices; however, more than 100 indices have been developed and applied to characterize them (Zargar, Sadiq, Naser, & Khan, 2011). Some drought indices can identify both wet and dry conditions, which may be able to identify flooding, for example, SPI and SPEI.

As the thesis started with an assumption that the SPEI can be used to identify both floods and droughts, it is therefore included (in addition to basic climatic variables) in the study of climate variability, trends, and their impacts on rice yields in (Chapter 3). The SPEI index at 1-month time scale (SPEI-1) demonstrates a stronger correlation with rice yields than monthly precipitation. Furthermore, yield impacts due to changes in SPEI-1 are clearer and point in the same direction with less uncertainty than those due to changes in minimum and maximum temperatures. However, the SPEI-1 cannot be used to identify areas that may be more prone to floods and droughts than others because of its inherent standardized nature. This standardization means that each level of drought severity classified by the SPEI (extreme wet, severe wet, moderate wet, moderate drought, severe drought, and extreme drought) occurs with the same frequency at all locations (Lloyd-Hughes & Saunders, 2002).

Based on the above description, and in an attempt to move forward and to obtain outcomes that cover all necessary aspects of floods and droughts associated with rice cultivation, the study decides to assess floods and droughts separately. For flooding, flood modeling is pivotal in forecasting and warning systems and can be used for identification of the areas liable to flooding (Popescu, Jonoski, Van Andel, Onyari, &

Moya Quiroga, 2010). Thus, an integrated hydrologic and hydraulic model of the Mun River is developed (Chapter 4).

For drought, a new drought risk assessment scheme is proposed for identification of areas susceptible to drought risk. The scheme combines three key elements: hazard, exposure, and vulnerability. Drought hazard is estimated by percentages of water deficit to rice cultivation, which, from the literature reviewed, appears to be a novel method. Exposed population and rice are the two elements of drought exposure, and vulnerability analysis consists of physical and socioeconomic factors as well as coping and adaptive capacity (Chapter 5).

7.1.2 Flood and drought impact on rice cultivation

Distinct natures and characteristics of floods and droughts challenge different opportunities for farmers' adaptive capacity to potential damage. The adverse consequences from both hazards on rice cultivation emerge from an imbalance between water supply and water demand brought about by rainfall variability (Krysanova et al., 2008). Excess water supply dominates in case of flooding. Floods in tropical countries can be highly variable. They can occur quickly and are intense; their duration can last a few hours or several consecutive days. In this case, apart from mitigation measures implemented in advance, there are fewer opportunities for people to do anything in order to prevent losses.

Whereas, droughts are a creeping phenomenon for which the impacts accumulate over time. This offers more rooms for farmers to adapt or change farm management practices so that they can reduce water demand and safeguard their rice and production. By lowering demand placed on water resources when droughts occur, their adverse effects will be diminished or disappeared. This is because droughts are a case of insufficient water supply which may arise from a deficiency in water supply, or an excess demand for water utilization, or both. This also represents that the farmers' coping ability to deal with droughts are evolving over time, even during drought events. This can partly be another reason of increasing trends of rice yields over the entire basin (Figure 7-1) although droughts periodically take place.

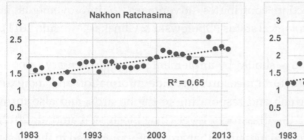 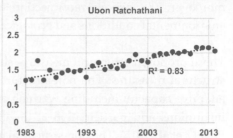

Figure 7-1 Trends of rice yields in Nakhon Ratchasima (in the west) and Ubon Ratchathani (in the east) provinces from 1984-2013 in the Mun River Basin in Thailand.

In addition, areas damaged by flooding are mostly clustered at low land or near rivers and tributaries, whereas those affected by droughts cover an extensive area. The areas affected by droughts can even be nearby rivers. Although these areas are generally less vulnerable, they commonly accommodate a lot of people, so the demand placed on water resources is relatively high compared to areas further away. As a result, droughts tend to cause more significant damage to people and agricultural regions than floods do.

7.1.3 Mitigation measures and policy development

Among mitigation measures already implemented in the basin, investment has mostly been placed in structural measures emphasizing augmentation of surface water storages (Chapter 6). With less attention being paid to other types of water storage, water demand reduction, and non-structural measures. This makes water storages less flexible and confines the coping capacity of farmers to only one measure.

If solutions from the range of mitigation measures and strategies are chosen wisely, not only can the problem of floods be solved, but significant benefits can be gained for drought protection, agricultural production and ecosystems functioning, etc. (Pavelic et al., 2015). Therefore, to solve flood and drought problems, integrated options are proposed to work in complementary with existing measures. They are selected based on four criteria which ensure the feasibility and practicability of the development projects (Chapter 6).

The proposed integrated mitigation measures involve surface and subsurface storages. The surface storages include improvement of existing water development projects, farm ponds, and oxbow lake reconnection. While the subsurface storage is a subsurface floodwater harvesting system. These proposed measures will make the storage system more flexible, and increase the overall storage capacity in the basin;

moreover, the oxbow reconnection contributes advantages to improvements in environmental conditions and ecological status (Obolewski & Glińska-Lewczuk, 2011).

Those proposed structural measures should be implemented complementary to non-structural measures. The non-structural measures provide information and increase adaptive capacity including warning and responding to the hazards, and also form institutions and participation at the grassroots level (Jha, Bloch, & Lamond, 2012; Pavelic et al., 2015). The non-structural measures that are briefly suggested in the thesis include field management techniques for rice cultivation and land use planning and policy (Chapter 6). These non-structural measures should be taken into serious consideration by the government, authorities, and all relevant stakeholders.

Field management techniques (furrow irrigation, alternate wetting and drying irrigation (AWD) and rice ratooning) diminish water demand for rice cultivation and increase water productivity. Numerous studies have documented these techniques in different countries. For example, furrow irrigation is proved to reduce water use by 40% and 31% and increase rice yields by 6% and 12-16% in Egypt and China, respectively, compared to the traditional flooded practice (Atta, 2008; Wang et al., 2018). The AWD increases rice production in China by 6 - 10% with 25% less water relative to the continuous-flooded regime (Wang et al., 2018); and AWD in the Philippines had the same yields as constant flooding but saved 11-20% in water use (Lampayan et al., 2004). Rice ratooning is the process of regenerating new tillers from the stubbles of a harvested main crop with minimal inputs. It allows farmers to gain 40-60% additional yields from second ratoon rice, using about 50-60% less labor and 60% less water. The ratoon rice takes only about half the time to mature (Elias, 1969; Oad, Sta Cruz, Memon, Oad, & Hassan, 2002; Santos, Fageria, & Prabhu, 2003).

Land-use planning minimizes development in flood-prone areas, designates routes and open spaces for better response and recovery efforts, and mitigates damages from unavoidable flood risk. Moreover, it conserves floodplains and wetlands as natural water storages that partially help in decreasing flood peaks and damage.

If all structural and non-structural measures are carefully studied and effectively implemented, this will lead to an improvement in resilience for adaptation to extreme climate events. Doing so will enhance agricultural productivity, ensure food security, help ameliorate poverty, and spur economic growth in the basin and of the country.

7.2 Main contributions & prospects of further research

The main contribution of including the SPEI-1 in the investigation of the past climatic conditions and their impacts on rice yields is that the SPEI-1 has a strong relationship with the yields. This is important, as the index can be used for drought monitoring and forecasting in many countries (Jeong, Sushama, & Naveed Khaliq, 2014; Labudová, Turňa, & Nejedlík, 2015; Turco et al., 2017; Xiao, Zhang, Singh, & Chen, 2017; Zappa et al., 2014). In addition, the trigger values for declaring drought and corresponding response plans should be studied and developed. This should be part of non-structural measures that will increase the coping ability of farmers to extreme climate events.

A contribution to drought risk analysis is the proposed scheme concept that combines all three key components (hazard, exposure, and vulnerability). Based on the literature reviewed, this is the first time that all the three key components are applied in quantitative drought risk assessment. However, the scheme is employed at the Mun River Basin in Thailand with the assumption that relative weights from each factor and each component to drought risk are equal. This assumption requires further studies regarding the relative weights of each element contributing to drought risk, and the weights of each factor contributing to each sub-component (exposed population, exposed rice, physical susceptibility, socioeconomic susceptibility, and coping and adaptive capacity), and each component.

For flooding, the main contributions are the first completed flood hazard maps at the Mun River Basin in Thailand, and the validated hydrologic and hydraulic models. The models can be further developed for the design of possible structural measures, and flood forecasting and warning systems. However, the main limitation of the thesis is data quality and availability, which influence the outcomes of the studies to a certain extent. In this case, the development of the hydraulic model is confined to the main river, even though there is also flood inundation at the tributaries. Thus, hydraulic models for all sub-basins should be developed to identify flood-prone areas upstream of the main channel.

Floods and drought can be monitored by means of detailed hydrologic models at the sub-basin scale. The hydrologic models of each sub-basins should, therefore, be created and linked with real-time gauging data. Subsequently, studies of water level thresholds or discharges triggering floods and droughts at each gauge station should also be carried out.

The last main contribution is the proposed mitigation measures for solving flood and drought problems in the basin. If these measures are carefully implemented, the two climatic problems may be overcome. That will ensure food availability, poverty reduction, and economic enhancement in the area. However, the proposed mitigation measures are selected from ones that proved successful when implemented in other places; they should be further verified by conducting pilot-projects in the study area, or by model simulation if there are available data. Also, appropriate non-structural measures that could be implemented complementary with the existing and proposed measures should be studied and verified as the proposed measures. Furthermore, a benefit-cost analysis taking into account societal, economic, and environmental impacts should be conducted.

References

Asian Disaster Preparedness Center (ADPC), & United Nations Development Programme (UNDP). (2005). *A primer: Integrated Flood Risk Management in Asia.* Bangkok, Thailand: Asian Disaster Preparedness Center.

Atta, Y. I. (2008). *Innovative Method for Rice Irrigation with High Potential of Water Saving.* Paper presented at the ICID Congress, integrated water resources management–from concepts to actions. , Lahore, Pakistan.

Centre for Research on the Epidemiology of Disaster (CRED). (2014). EM-DAT: The International Disaster Database. from Centre for Research on the Epidemiology of Disaster - CRED, Université Catholique de Louvain, Brussels, Belgium www.emdat.be

Elias, R. (1969). Rice production and minimum tillage. *Outlook on Agriculture, 6*(2), 67-71.

Jeong, D. I., Sushama, L., & Naveed Khaliq, M. (2014). The role of temperature in drought projections over North America. *Climatic Change, 127*(2), 289-303. doi:10.1007/s10584-014-1248-3

Jha, A. K., Bloch, R., & Lamond, J. (2012). *Cities and flooding: a guide to integrated urban flood risk management for the 21st century*: The World Bank.

Krysanova, V., Buiteveld, H., Haase, D., Hattermann, F., Van Niekerk, K., Roest, K., . . . Schlüter, M. (2008). Practices and lessons learned in coping with climatic hazards at the river-basin scale: floods and droughts. *Ecology and Society, 13*(2).

Labudová, L., Turňa, M., & Nejedlík, P. (2015). *Drought monitoring in Slovakia.* Paper presented at the Towards Climatic Services. International scientific conference.

Lampayan, R., Bouman, B. A., De Dios, J., Lactaoen, A., Espiritu, A., Norte, T., . . . Soriano, J. (2004). *Adoption of water saving technologies in rice production in the Philippines*: Food & Fertilizer Technology Center.

Lloyd-Hughes, B., & Saunders, M. A. (2002). A drought climatology for Europe. *International Journal of Climatology, 22*(13), 1571-1592.

Oad, F., Sta Cruz, P., Memon, N., Oad, N., & Hassan, Z. U. (2002). Rice ratooning management. *Journal of Applied Sciences, 2*, 29-35.

Obolewski, K., & Glińska-Lewczuk, K. (2011). Effects of oxbow reconnection based on the distribution and structure of benthic macroinvertebrates. *Clean–soil, air, water, 39*(9), 853-862.

OECD. (2016). Mitigating Droughts and Floods in Agriculture: Policy Lessons and Approaches. *OECE Studies on Water,* 72. doi:https://doi.org/10.1787/9789264246744-en

Pavelic, P., Brindha, K., Amarnath, G., Eriyagama, N., Muthuwatta, L., Smakhtin, V., . . . Sharma, B. R. (2015). *Controlling floods and droughts through underground storage: from concept to pilot implementation in the Ganges River Basin* (Vol. 165): International Water Management Institute (IWMI).

Pavelic, P., Srisuk, K., Saraphirom, P., Nadee, S., Pholkern, K., Chusanathas, S., . . . Smakhtin, V. (2012). Balancing-out floods and droughts: opportunities to utilize floodwater harvesting and groundwater storage for agricultural development in Thailand. *Journal of Hydrology, 470,* 55-64.

Popescu, I., Jonoski, A., Van Andel, S., Onyari, E., & Moya Quiroga, V. (2010). Integrated modelling for flood risk mitigation in Romania: case study of the Timis–Bega river basin. *International journal of river basin management, 8*(3-4), 269-280.

Santos, A., Fageria, N., & Prabhu, A. (2003). Rice ratooning management practices for higher yields. *Communications in soil science and plant analysis, 34*(5-6), 881-918. doi: 10.1081/CSS-120018981

Turco, M., Ceglar, A., Prodhomme, C., Soret, A., Toreti, A., & Francisco, J. D.-R. (2017). Summer drought predictability over Europe: empirical versus dynamical forecasts. *Environmental research letters, 12*(8), 084006.

United Nations Environment Programme (UNEP), & International Environmental Technology Centre (IETC). (1998). *Source book of alternative technologies for freshwater augmentation in some countries in Asia (Technical Publication)* (Vol. no. 8B). Osaka, Japan: UNEP International Environmental Technology Centre.

Wang, Z., Gu, D., Beebout, S. S., Zhang, H., Liu, L., Yang, J., & Zhang, J. (2018). Effect of irrigation regime on grain yield, water productivity, and methane emissions in dry direct-seeded rice grown in raised beds with wheat straw incorporation. *The Crop Journal, 6*(5), 495-508.

Wilhite, D. A., & Vanyarkho, O. (2000). Drought: Pervasive impacts of a creeping phenomenon. *Drought: A global assessment, 1,* 245-255.

Xiao, M., Zhang, Q., Singh, V. P., & Chen, X. (2017). Probabilistic forecasting of seasonal drought behaviors in the Huai River basin, China. *Theoretical and applied climatology, 128*(3-4), 667-677.

Zappa, M., Bernhard, L., Spirig, C., Pfaundler, M., Stahl, K., Kruse, S., . . . Stähli, M. (2014). A prototype platform for water resources monitoring and early recognition of critical droughts in Switzerland. *Proceedings of the International Association of Hydrological Sciences, 364,* 492-498. doi:10.5194/piahs-364-492-2014

Zargar, A., Sadiq, R., Naser, B., & Khan, F. I. (2011). A review of drought indices. *Environmental Reviews, 19*(NA), 333-349. doi:10.1139/a11-013

About the Author

Saowanit Prabnakorn was born in Bangkok, Thailand. She earned a Bachelor degree in Civil Engineering from Mahidol University in Thailand in 2002. In 2005, she continued her Master degree at National Institute of Development Administration (NIDA), Thailand in parallel with working as a highway & drainage engineer at a private company. Afer she obtained her Master degree in Business Economics in 2007, she changed her career to a financial field, working as a credit review & risk officer. Two years after, she became a lecturer at the Department of Business Economics, Faculty of Liberal Art and Management Science, Prince of Songkhla University (Surat Thani Campus) in the south of Thailand. In 2012, she won the opportunities to pursue another master study and continue with a Ph.D. in the Netherlands by a full scholarship from the Royal Thai Government. With this scholarship, she has turned back to Engineering disciplines and started her long journey. She obtained her second Master degree (with distinction) in Hydraulic Engineering (Land & Water Development) from (was) UNESCO-IHE, Delft, the Netherlands on April 2014. Immediately, she has continued her Ph.D. in the same specialization, and has published several research articles in scientific journals and has presented her works in international conferences.

List of Publications

Journal Articles

Prabnakorn, S., Maskey, S., Suryadi, F., & de Fraiture, C. (2018). Rice yield in response to climate trends and drought index in the Mun River Basin, Thailand. *Science of the Total Environment*, 621, 108-119. doi: 10.1016/j.scitotenv.2017.11.136.

Prabnakorn, S., Maskey, S., Suryadi, F., & de Fraiture, C. (2019). Assessment of drought hazard, exposure, vulnerability, and risk for rice cultivation in the Mun River Basin in Thailand. *Natural Hazards*, 97, 891-911. doi:10.1007/s11069-019-03681-6.

Prabnakorn, S., Suryadi, F., Chongwilaikasem, J., & de Fraiture, C. (2019). Development of an integrated flood hazard assessment model for a complex river system: a case study of the Mun River Basin, Thailand. *Modeling Earth Systems and Environment*, doi:10.1007/s40808-019-00634-7.

Prabnakorn, S., Ruangpan, L., Suryadi, F., & de Fraiture, C. (2019). Solving floods and droughts for rice cultivation at the Mun River Basin in Thailand. *(Submitted)*

Conference presentations

Prabnakorn, S., Suryadi, F. X., & de Fraiture, C. (2015, 12-17 April 2015). *Flood Risk Assessment as a Part of Integrated Flood and Drought Analysis. Case Study: Southern Thailand.* Paper presented at the EGU General Assembly Conference Abstracts, Vienna, Austria.

Prabnakorn, S., Maskey, S., Suryadi, F., & de Fraiture, C. (2016, 6-8 Nov 2016). *Climate and Drought Trends and Their Relationships with Rice Production in the Mun River Basin, Thailand.* Paper presented at the International Commission on Irrigation and Drainage (ICID) Conference Proceedings, Chiang Mai, Thailand.

Prabnakorn, S., Maskey, S., Suryadi, F., & de Fraiture, C. (2018, 12 - 17 August 2018). *Drought hazard to rice cultivation in the Mun River Basin, Thailand.* Paper presented at the International Commission on Irrigation and Drainage (ICID) Conference Abstracts, Saskatchewan, Canada.

Acknowledgment

First, I would like to thank my esteemed professor Prof. Dr. C. de Fraiture for her contribution to me. I could say that I always gain new perspectives from every discussion, and that navigate and drive me through my academic work until I accomplish. It is a great honor to study and work with you.

I am deeply indebted to Dr. F.X. Suryadi for his patience, open-minded, generosity, suggestions and support thought out my tough period, especially when I am sometimes persistent, he always accepts and gives me the opportunities and freedom to follow my ideas and learn from my experiences.

My gratitude is conveyed to the authorities of the Royal Irrigation Department, Department of Water Resources, the Meteorological Department, Land Development Department, and Office of Agricultural Economics in Thailand for providing data and information for my thesis as well as the advice and assistance to facilitate my work.

I am much obliged to Head of the Department of Agricultural Engineering, Faculty of Engineering, Rajamangala University of Technology Thanyaburi, Thirapong Kuankhamnuan, for the support.

I truly appreciate all the support, encouragement, attention, motivation, and kindness from all my friends in IHE and my Thai friends here and abroad. Though I do not mention your names, the memorable experiences between us will always be in my hearts; no one is missing.

Above all, my sincere gratitude goes to my mother, brother, husband and son, and everyone in my family, who are always beside me, understanding me and giving me unconditional love, support, and encouragement. Another person who is always in my heart is my father, my first teacher who dedicated his life for the others' happiness. "Papa, without your inspiration, I could not be as I am now. Thank you very much for everything you were giving me".

Last but not least, I wish to acknowledge the financial support from the Royal Thai Government to give me the opportunity to study in IHE Delft, Institute for Water Education, Delft, Netherlands.

Netherlands Research School for the
Socio-Economic and Natural Sciences of the Environment

DIPLOMA

For specialised PhD training

The Netherlands Research School for the
Socio-Economic and Natural Sciences of the Environment
(SENSE) declares that

Saowanit Prabnakorn

born on 8 September 1979, Bangkok, Thailand

has successfully fulfilled all requirements of the
Educational Programme of SENSE.

Delft, 22 January 2020

The Chairman of the SENSE board

Prof. dr. Martin Wassen

the SENSE Director of Education

Dr. Ad van Dommelen

The SENSE Research School has been accredited by the Royal Netherlands Academy of Arts and Sciences (KNAW)

KONINKLIJKE NEDERLANDSE
AKADEMIE VAN WETENSCHAPPEN

The SENSE Research School declares that Saowanit Prabnakorn has successfully fulfilled all
requirements of the Educational PhD Programme of SENSE with a
work load of 38.9 EC, including the following activities:

SENSE PhD Courses

o   Environmental research in context (2014)
o   Research in context activity: 'Co-organizing and contributing to course on: Managing
    flood for spate irrigation development related to food crops security reflex to the impact
    of climate changes (Bandung, Indonesia on 26-30 October 2015)'

Other PhD and Advanced MSc Courses

o   Where there is little data: How to estimate design variable in poorly gauged basins, IHE
    Delft (2014)
o   Conveyance and irrigation structure, IHE Delft (2018)

Management and Didactic Skills Training

o   Teaching in the MSc course 'AquaCROP model for achieving productivity under water
    scarcity' (2014)
o   Teaching in the MSc course 'Irrigation & drainage main system design' (2015-2017)
o   Co-organizing and team coordination IHE PhD symposium and symposium overview
    booklet (1-2 October 2018)

Oral Presentations

o   *Integrated flood and drought mitigation measures and strategies. Case study: The Mun
    River Basin, Thailand.* IHE PhD symposium, 28-29 September 2015, Delft, The
    Netherlands
o   *Historical climate and drought variability and their relationships with rice yield. Case
    Study: The Mun River Basin, Thailand.* IHE PhD symposium, 3-4v November 2016, Delft,
    The Netherlands
o   *Climate and drought trends and their relationships with rice production in the Mun River
    Basin, Thailand.* 2nd World Irrigation Forum (WIF2), International Commission on
    Irrigation and Drainage, 6-8 November 2016, Chiang Mai, Thailand
o   *Drought hazard to rice cultivation in the Mun River Basin, Thailand.* International
    Conference and 69th International Executive Council Meeting of the International
    Commission on Irrigation and Drainage, 12-17 November, Saskatoon, Canada

SENSE Coordinator PhD Education

Dr. ir. Peter Vermeulen